ANDERSONIAN LIBRARY

25 MAY

BOUND

Bridge Deck Behaviour

EDMUND C. HAMBLY

LONDON
CHAPMAN AND HALL

A Halsted Press Book
JOHN WILEY & SONS, INC., NEW YORK

First published 1976
by Chapman and Hall Ltd
11 New Fetter Lane, London EC4P 4EE
© 1976 E. C. Hambly
Typeset by Preface Ltd, Salisbury, Wilts
and printed in Great Britain by
T. & A. Constable Ltd, Edinburgh

ISBN 0 412 13190 0

All rights reserved. No part of this book may
be reprinted, or reproduced or utilized in any form or
by any electronic, mechanical or other means, now known
or hereafter invented, including photocopying and recording,
or in any information, storage or retrieval system,
without permission in writing from the publisher.

Distributed in the U.S.A.
by Halsted Press, a Division
of John Wiley & Sons, Inc., New York

Library of Congress Cataloging in Publication Data

Hambly, Edmund C
    Bridge deck behaviour.

    Includes bibliographical references and indexes.
    1. Bridges — Floors. I. Title.
TG325.6.H35   624.2'52   75-16496
ISBN 0-470-34636-1

D
624.253
HAM

To the inspiring memory of
Kenneth H. Roscoe and Stuart G. Spickett

Acknowledgements

This book developed from numerous discussions I have had with colleagues and friends. My interest in the subject developed while I worked under Robert Benaim at Ove Arup and Partners; later John Blanchard and Peter Dunican encouraged me to write a book. Most of my practical experience in bridge design and analysis was gained while I worked as a Senior Engineer for E. W. H. Gifford and Partners under Maurice Porter. I am grateful to them for this experience and for paying some of the costs of preparing the manuscript. During this period Ernest Pennells was a most stimulating colleague. The manuscript was read by Dr J. E. Spindel and R. Benaim. I am grateful for their numerous constructive criticisms. I should also like to thank the publisher's technical advisor for his advice.

Finally, I must mention that the writing of this book was very much a joint effort by my wife and I. Her contribution was as significant and demanding as mine, and I thank her wholeheartedly for being such a cooperative and encouraging partner.

E.C.H.

Contents

	page
Preface	xi
Notation	xiv

1. Structural forms of bridge decks   1
   1.1 Introduction   1
   1.2 Structural forms   4

2. Beam decks: continuous beam analysis   19
   2.1 Introduction   19
   2.2 Types of structure   19
   2.3 Bending of beams   21
   2.4 Torsion of beams   36
   2.5 Computer analysis of continuous beams   42
   2.6 Construction sequence   43
   2.7 Redistribution of moments   45
   2.8 Beam decks with raking piers   45
   References   45

3. Slab decks: grillage analysis — 46
- 3.1 Introduction — 46
- 3.2 Types of structure — 47
- 3.3 Structural action — 48
- 3.4 Rigorous analysis of distribution of forces — 55
- 3.5 Grillage analysis — 55
- 3.6 Grillage examples — 63
- 3.7 Interpretation of output — 67
- 3.8 Moments under concentrated loads — 68
- References — 68

4. Beam-and-slab decks: grillage analysis — 70
- 4.1 Introduction — 70
- 4.2 Types of structure — 70
- 4.3 Structural action — 71
- 4.4 Grillage analysis — 74
- 4.5 Grillage examples — 77
- 4.6 Application of load — 81
- 4.7 Interpretation of output — 82
- 4.8 Slab membrane action in beam-and-slab decks — 83
- References — 85

5. Cellular decks: shear-flexible grillage analysis — 87
- 5.1 Introduction — 87
- 5.2 Types of structure — 87
- 5.3 Grillage mesh — 88
- 5.4 Modes of structural action — 89
- 5.5 Section properties of grillage members — 103
- 5.6 Load application — 108
- 5.7 Interpretation of output — 109
- 5.8 Other methods of analysis — 113
- References — 115

6. Shear key decks — 116
- 6.1 Introduction — 116
- 6.2 Structural behaviour — 116
- 6.3 Grillage analysis — 119
- 6.4 Skew decks — 124
- References — 125

7. Three-dimensional space frame analyses and slab membrane action — 126
   7.1 Slab membrane action — 126
   7.2 Downstand grillage — 127
   7.3 McHenry lattice space frame — 134
   7.4 Cruciform space frame — 137
   7.5 Effects of slab membrane action on beam-and-slab deck behaviour — 139
   References — 141

8. Shear lag and edge stiffening — 142
   8.1 Shear lag — 142
   8.2 Effective width of flanges — 143
   8.3 Edge stiffening of slab decks — 146
   8.4 Upstand parapets to beam-and-slab decks — 148
   8.5 Service bays to beam-and-slab decks — 150
   References — 152

9. Skew, tapered and curved decks — 153
   9.1 Skew decks — 153
   9.2 Tapered decks — 158
   9.3 Curved decks — 159
   References — 162

10. Charts for preliminary design — 163
   10.1 Introduction — 163
   10.2 Some published load distribution charts — 164
   10.3 Influence lines for slab, beam-and-slab and cellular decks — 166
   10.4 Application of charts to slab deck — 173
   10.5 Application of charts to beam-and-slab deck — 177
   10.6 Application of charts to cellular deck — 181
   References — 185

11. Temperature and prestress loading — 187
   11.1 Introduction — 187
   11.2 Temperature strains and stresses in simply supported span — 187
   11.3 Temperature stresses in continuous deck — 192
   11.4 Grillage analysis of temperature moments — 195
   11.5 Differential creep and shrinkage — 196
   11.6 Prestress axial compression — 196

11.7 Prestress moments due to cable eccentricity	197
11.8 Prestress moments due to cable curvature	198
11.9 Prestress analysis by flexibility coefficients	201
References	203

12. Harmonic analysis and folded plate theory — 204

12.1 Introduction	204
12.2 Harmonic components of load, moment, etc.	205
12.3 Characteristics of low and high harmonics	210
12.4 Harmonic analysis of plane decks	214
12.5 Folded plate analysis	216
12.6 Continuous and skew decks	221
12.7 Errors of harmonics near discontinuities	222
References	223

13. Finite element method — 224

13.1 Introduction	224
13.2 Two-dimensional plane stress elements	225
13.3 Plate bending elements	233
13.4 Three dimensional plate structures and shell elements	237
13.5 Finite strips	238
13.6 Three-dimensional elements	241
13.7 Conclusion	241
References	242

Appendix A.
Product integrals. Functions of load on a single span. Harmonic components — 243

Appendix B.
Approximate folded plate method for beam-and-slab and cellular decks — 247

Author Index — **269**
Subject Index — **270**

Preface

This book is for use in bridge design offices as a guide to methods of bridge deck analysis. It has been written to be intelligible to junior engineers who are interested in the general physical characteristics of the different types of bridge deck and who also require detailed descriptions of the techniques most commonly used. The book has also been written with consideration for the senior engineers leading design teams to give them information of the variety and appropriateness of analytical methods available and their various shortcomings.

With the great simplification in computer methods in recent years it has become possible for engineers to analyse complex structures without complex or cumbersome mathematics. Since the majority of design office engineers only have confidence in their calculations when they can back them up with physical reasoning, the book concentrates on the physical reasoning that is necessary to translate prototype behaviour and properties into computer models, and vice versa. Most attention is paid to the simpler models of continuous beam and grillage because they are most commonly used. With experience engineers are able to use physical reasoning and simple models for the design of relatively complex structures. But since such experience involves comparisons of results of these simple methods with test results and solutions of more rigorous analyses, the principles of space frame, folded plate and finite element methods are described in later chapters. The only mathematics that are necessary for the majority of bridge deck designs are summarized in Chapter 2 and concern simple beam theory that is covered in most university first year courses on civil and structural engineering.

Hand methods of analysis are also very useful and will remain essential for preliminary design, checks, and when the computer is not available. Initially the author greatly preferred such methods to the general use of the computer, and an early draft of this book concentrated on the subject. However with increased experience and responsibility, a complete change of attitude became necessary because the computer methods had the advantage of being:

(1) Comprehensible to the majority of engineers, many of whom, though thoroughly competent, do not have the mathematical expertise in techniques, such as harmonic analysis, that are needed for accurate application of many hand methods to complicated structures.
(2) Applicable to the majority of bridge shapes with skew, curved or continuous decks and with varying stiffness from region to region. In contrast, hand methods are only simple to use for the few bridges which are rectangular in plan and simply supported.
(3) Applicable, with shear flexibility, to a much wider variety of deck cross-sections.
(4) Checkable. It is much easier to check computer data and output distributions of forces than pages of hand calculations.
(5) Economical. With the development of very convenient and clear grillage programs computer data can be prepared, numerous load cases analysed, and the results processed in a much shorter time than the equivalent hand calculations can be carried out.

Nonetheless, because hand methods are still very useful, some published techniques are reviewed in Chapter 10 and applications of rapid design charts are demonstrated.

Live loading has not been discussed with regard to influence lines because such loading often depends on the particular client's requirements and interpretation of guidelines. However, Chapter 11 describes how temperature and prestress influence a structure, and how such loading is simulated in the analytical models. Temperature loading has been considered in rather more detail in recent years. It is often found that calculated stresses due to temperature exceed those due to live load.

It has been assumed throughout the book that bridges behave as linearly elastic structures. Plastic and yield line methods are not described because though increasing in popularity they are not generally used. Since superposition of load cases is not applicable they cannot easily be used with load distribution methods. Furthermore, they have not yet been developed sufficiently for the analysis of the more complex structures.

The accuracy of any method of analysis for a particular structure is difficult to predict or even check. It depends on the ability of the model to represent three very complex characters: the behaviour of the material, the geometry of the structure, and the actual loading. Construction materials, even when homogeneous, have properties differing widely from the elastic, or plastic, idealizations. When incorporated in a structure they have innumerable variations of stiffness and strength due to composition and site- and life-histories. The analysis almost invariably simplifies the geometry of the structure of thick members to an assemblage of thin plates or beams. Numerous holes, construction joints, site imperfections and other details are ignored. Finally the design loadings for live load, temperature, creep, settlement and so on are idealizations based on statistical studies. It is unlikely that the critical design load will ever act on the structure even though it might be exceeded. For these reasons, large errors are likely whatever method of analysis is used. It is suggested that greater emphasis should be given to considering the physical behaviour of the structure and anticipating consequences of calculations being in error by more than 10 per cent than to refining calculations in pursuit of the last 1 per cent of apparent accuracy.

E. C. Hambly
January, 1975

Notation

Superscripts

−	average value or global variable
ˆ	maximum value
′	relates to top slab of cellular deck
″	relates to bottom slab of cellular deck

Subscripts

c	of centroid
e	of equivalent grillage member, or of effective flange
f	of flange
l	longitudinal
M	due to bending
S	due to shear
T	due to torsion
t	transverse
W	related to loading
w	of web
x, y, z	axis of member, or moment, force or section property related to vertical bending of that member

xx, yy, zz   local axis for direction of force and associated shear area or about which moment acts
I, II   principal values
1, 2, ... n number of, member end, or slab edge, or support, or node, or beam, or harmonic

A	area of, cross section, or part section, or enclosed area
A_S	equivalent shear area
a	stiffness coefficient, or dimension, or harmonic coefficient
a_S	equivalent shear area per unit width
b	breadth, or stiffness coefficient, or harmonic coefficient
C	torsion constant
c	torsion constant per unit width, or stiffness coefficient
c	cellular stiffness ratio
D	flexural rigidity
d	depth, or thickness
E	Young's Modulus
e	eccentricity of prestress
F	node force
f	flexural stiffness ratio
G	Shear Modulus
g	stiffness coefficient
h	distance between midplanes of slabs in cellular decks
h', h''	distances of midplanes of top and bottom slabs from their common centroid
I	moment of inertia = second moment of area
i	moment of inertia per unit width
J	influence value
j	shear flexibility parameter
K, k	stiffness matrices
k	spring stiffness, or stiffness coefficient
L	span, or distance between points of contraflexure
l	length, or web, or 'beam' spacing
M	bending moment
M_{FE}	fixed end moment
m	moment per unit width, or modular ratio, or moment system in flexibility analysis
N	applied torque
n	harmonic number, or stiffness coefficient

O	origin
P	axial tension force, or prestress compression force
R	radius of curvature, or reaction
r	force matrices
r	shear flow
r	rotational stiffness ratio
S	shear force
S_{FE}	fixed end shear force
s	shear force per unit width, or distance around curved arc or midplane
T	torque
t	torque per unit width, or thickness
t	axis rotation transformation matrix
U	applied load in Ox direction
u	displacement in Ox direction (warping)
U, u	force or displacement matrices
V	applied load in Oy direction
v	displacement in Oy direction
W	applied load in Oz direction (vertical downwards)
w	displacement in Oz direction (vertical downwards)
Ox	horizontal axis along span (except where given local direction)
X, **x**	release action in flexibility analysis
X	load matrix
Oy	horizontal axis transverse to span (except where given local direction)
y	horizontal distance of point to side of origin or neutral axis
Z	amplitude of harmonic component of vertical load
Oz	vertical axis downwards (except where given local direction)
z	vertical distance of point below origin or neutral axis
α	angle, or coefficient or thermal expansion, or $(n\pi/L)$
$\alpha_1, \alpha_2 \ldots$	coefficients of displacement field
γ	shear strain
Δ	increment
δ	flexibility coefficient
ϵ	linear strain
θ	rotation, usually slope $\partial w/\partial x$, or inclination of prestress
ν	Poisson's Ratio
σ	tension/compression stress
τ	shear stress
ϕ	rotation, usually $\partial w/\partial y$
$\dot{\phi}$	twist = rate of change of ϕ with length

1
Structural forms of bridge decks

1.1 Introduction

Bridge decks are developing today as fast as they have at any time since the beginning of the Industrial Revolution. While few developments can be as outstanding as Robert Stephenson's construction of the Britannia Bridge, the diversity of sites is increasingly challenging the ingenuity of engineers to produce new structural forms and appropriate materials. Methods of analysis have developed simultaneously and in the last thirty years progress has been particularly significant. Hand methods of load distribution and more recently the application of digital computers have enabled engineers to analyse decks with complex cross-sections and complicated skew, curved and continuous spans. These design methods demand a considerable amount of theoretical and experimental research to improve and test their reliability. However, several have now been developed to such usable form that with an understanding of physical behaviour designers can analyse complex decks without recourse to complicated mathematical theory.

This chapter reviews and categorizes the principal types of bridge deck construction currently being used, and refers to the analytical techniques demonstrated in later chapters. The types of construction are divided into beam, grid, slab, beam-and-slab and cellular to differentiate their individual geometric

Fig. 1.1 Britannia Bridge, Wales (1849). Wrought iron box girder deck with a 140 m (460 ft) main span. Designed by Robert Stephenson. From a lithograph by G. Hawkins, 1850. Courtesy of The Institution of Civil Engineers, London.

Structural Forms of Bridge Deck 3

Fig. 1.2 Detail of Britannia Bridge. Courtesy of the Science Museum, London SW7.

4   Bridge Deck Behaviour

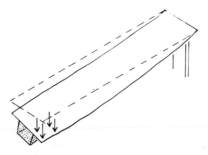

Fig. 1.3   Beam deck bending and twisting without change of cross-section shape.

and behavioural characteristics. Inevitably many decks fall into more than one category, however they can usually be analysed by using a judicious combination of the methods applicable to the different types.

1.2 Structural forms

1.2.1 Beam decks

A bridge deck can be considered to behave as a beam when its length exceeds its width by such an amount that when loads cause it to bend and twist along its length, its cross-sections displace bodily and do not change shape.

The most common beam decks are footbridges, either of steel, reinforced concrete or prestressed concrete. They are often continuous over two or more spans. In addition some of the largest box-girder decks can be analysed as beams to determine the distributions of longitudinal bending moments, shears and torsions. In the Britannia Bridge and modern long span box-girders, the dominant load is concentric so that the distortion of the cross-section under eccentric loads has relatively little influence on the principal bending stresses.

The analysis of bending moments and torsions in continuous beam decks is discussed in Chapter 2.

1.2.2 Grid decks

The primary structural member of a grid deck is a grid of two or more longitudinal beams with transverse beams (or diaphragms) supporting the running slab. Loads are distributed between the main longitudinal beams by the bending and twisting of the transverse beams. Because of the amount of workmanship needed to fabricate or shutter the transverse beams, this method

Structural Forms of Bridge Deck 5

Fig. 1.4 Footbridge, Norway. Beam deck. Designed by Dr Ing. A. Aas-Jakobsen, engineer and Jonas Haanshus, architect. Courtesy of the Cement and Concrete Association, London.

6 *Bridge Deck Behaviour*

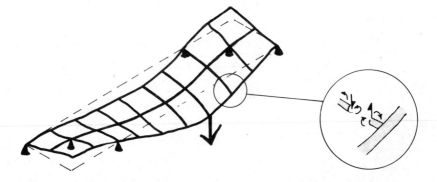

Fig. 1.5 Load distribution in grid deck by bending and torsion of beam members.

of construction is becoming less popular and is being replaced by slab and beam-and-slab decks with no transverse diaphragms.

Grid decks are most conveniently analysed with the conventional computer grillage analysis described in Chapter 3. The analysis in effect sets out a set of simultaneous slope-deflection equations for the moments and torsions in the beams at each joint and then solves the equations for the load cases required. Grid decks have also been analysed for many years by hand methods which are summarized in Chapter 10. However, such methods are decreasing in popularity as computer methods become simpler and more versatile.

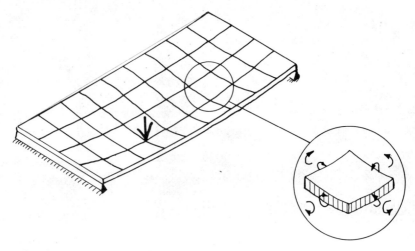

Fig. 1.6 Load distribution in slab deck by bending and torsion in two directions.

1.2.3 Slab decks

A slab deck behaves like a flat plate which is structurally continuous for the transfer of moments and torsions in all directions within the plane of the plate. When a load is placed on part of a slab, the slab deflects locally in a 'dish' causing a two-dimensional system of moments and torsions which transfer and share the load to neighbouring parts of the deck which are less severely loaded. A slab is 'isotropic' when its stiffnesses are the same in all directions in the plane of the slab. It is 'orthotropic' when the stiffnesses are different in two directions at right angles.

Concrete slab decks are commonly used where the spans are less than 15 m (50 ft). If the deck is cast *in situ* it is general practice to consider it as isotropic, even though the reinforcement may not be the same in all directions.

When it is inconvenient to support the deck on falsework during construction, the slab is often built compositely with reinforced concrete cast *in situ* between previously erected beams. The beams of precast concrete or steel have a greater stiffness longitudinally than the *in situ* concrete has transversely; thus the deck is orthotropic.

For spans greater than 15 m (50 ft), the material content and dead load of a solid slab becomes excessive and it is customary to lighten the structure by incorporating voids of cylindrical or rectangular cross-section near the neutral axis. If the depth and width of the voids is less than 60 per cent of the overall structural depth, their effect on the stiffness is small and the deck behaves effectively as a plate. Voided slab decks are frequently constructed of concrete cast *in situ* with permanent void formers, or of precast prestressed concrete box beams post-tensioned transversely to ensure transverse continuity. If the void size exceeds 60 per cent of the depth the deck is generally considered to be of cellular construction with a different behaviour, as is described later.

Rigorous analysis of most slab decks is not possible at present. However, slab decks can be conveniently analysed using the computer grillage as described in Chapter 3. In this method the continuous slab is represented by an equivalent grid of beams whose longitudinal and transverse stiffnesses are approximately the same as the local plate stiffnesses of the slab. This analogy is not the closest available, but it has been found to agree well with more rigorous solutions and is generally accepted as sufficiently accurate for most designs. One of the more exact methods, which is mentioned in Chapter 13, is performed using a computer finite element program. In this method the deck is notionally split up into a number of elements, frequently triangular, for each of which approximate plate bending equations can be derived and solved (as opposed to the beam equations used in the grillage). The method is much more complicated and

8 Bridge Deck Behaviour

expensive than the grillage, and since for slab decks it does not generally produce significantly different results, the grillage is normally used in preference. The hand methods of Chapter 10 also provide a convenient method of analysis for the design of slab decks with simple plan geometry. However, as mentioned for grid decks, computer methods are becoming easier to use and provide more information.

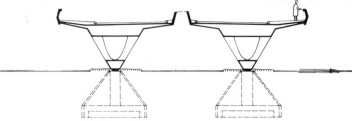

Fig. 1.7 Western Bank Bridge, Sheffield (1969). Slab deck. Designed by Ove Arup and Partners, London. Courtesy of Ove Arup and Partners.

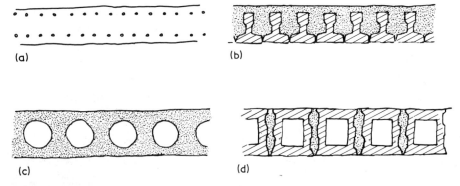

Fig. 1.8 Slab decks. (a) solid (b) composite of *in situ* concrete infilling precast beams (c) voided (d) voided of precast box beams post-tensioned transversely.

One type of deck which does not fit neatly into any of the main categories is the 'shear-key' deck. A shear key deck is constructed of contiguous prestressed/ reinforced concrete beams of rectangular or box sections, connected along their length by *in situ* concrete joints. It is not prestressed transversely and thus is not fully continuous for transverse moments. The main application is for bridges constructed over busy roads or railways, where it is convenient to erect the beams overnight and then complete the jointing without disturbance to the traffic below. Although such decks have little or no transverse bending stiffness, distribution of loads between beams still takes place because differential deflection of the beams is resisted by the torsional stiffness of the beams and a vertical shear force is transferred across the keyed joints. The analysis of such decks is described in Chapter 6.

1.2.4 Beam-and-slab decks

A beam-and-slab deck consists of a number of longitudinal beams connected across their tops by a thin continuous structural slab. In transfer of the load

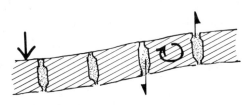

Fig. 1.9 Differential deflections of beams in shear key deck resisted by torsion of beams.

longitudinally to the supports, the slab acts in concert with the beams as their top flanges. At the same time the greater deflection of the most heavily loaded beams bends the slab transversely so that it transfers and shares out the load to the neighbouring beams. Sometimes this transverse distribution of load is assisted by a number of transverse diaphragms at points along the span, so that deck behaviour is more similar to that of a grid deck. However, the use of diaphragms is becoming less popular because of the construction problems they

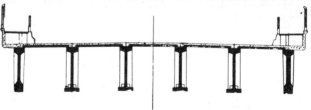

Fig. 1.10 Tasman Bridge, Australia (1964). Beam-and-slab deck designed by E. W. H. Gifford & Partners for G. Maunsell & Partners. Courtesy of E. W. H. Gifford & Partners.

cause and because their localized stiffnesses attract forces which can cause unnecessary stress concentrations. Beam-and-slab construction has the advantage over slab that it is very much lighter while retaining the necessary longitudinal stiffness. Consequently it is suitable for a much wider range of spans, and it lends itself to precast and prefabricated construction. Occasionally the transverse flexibility can be advantageous; for example, it can help a deck on skew supports to deflect and twist 'comfortably' under load without excessively loading the nearest supports to the load or lifting off those further away.

Beam-and-slab decks can be divided into two main groups: those with the beams at close centres or touching are referred to as 'contiguous beam-and-slab', while those with beams at wide centres are referred to as 'spaced beam-and-slab'. The most common form of contiguous beam-and-slab comprise precast prestressed concrete inverted T-beams supporting a cast *in situ* reinforced concrete slab of about 200 mm (8 in) thickness. When a load is placed on part of such a deck, the slab deflects in a smooth wave so that for load distribution its behaviour can be considered similar to that of an orthotropic slab with longitudinal stiffening.

Spaced beam-and-slab decks commonly have the beams at about 2 m (6 ft) to 3.5 m (12 ft) centres. Decks have been designed with precast prestressed concrete beams or steel beams supporting a concrete slab. Numerous variations of construction have been employed for the concrete slab ranging from totally cast *in situ* to very large precast panels connected to the beams by the minimum of *in situ* concrete in the joints. Sometimes the running slab on steel beams consists of a steel 'battledeck' which is constructed of a stiffened steel plate of as little as 12 mm (½ in) thickness. When a load is placed above one beam of a spaced beam-and-slab deck, the slab does not necessarily deflect transversely in a single wave but sometimes in a series of waves between beams. This is particularly the

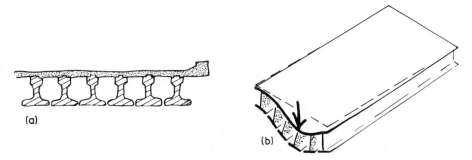

Fig. 1.11 (a) Contiguous beam-and-slab deck and (b) slab of contiguous beam-and-slab deck deflecting in smooth wave.

12 *Bridge Deck Behaviour*

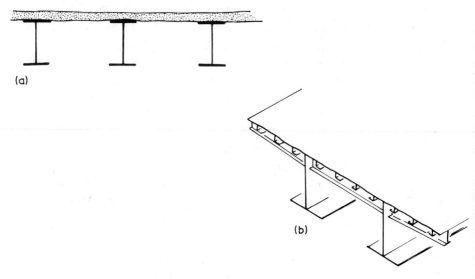

Fig. 1.12 Spaced beam-and-slab decks with steel I-beams and (a) concrete slab (b) steel 'battledeck' running slab.

case if the beams have high torsional stiffnesses, as do box beams, when the beams may twist little so that the slab deflects in a series of transverse steps. Such behaviour is different from that of orthotropic slabs and it is advisable for the analytical model to have its high longitudinal beam stiffnesses correctly positioned across the deck. If the various stiffnesses cannot be correctly located, the load distribution analysis is sometimes backed up with a plane frame analysis of the cross-section to study its transverse bending and distortion.

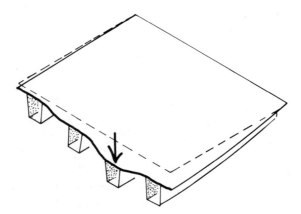

Fig. 1.13 Spaced beam-and-slab deck deflecting in series of steps or waves.

Structural Forms of Bridge Deck   13

The extreme form of spaced beam-and-slab decks can have as few as two spine beams at more than 12 m (40 ft) centres. Solid concrete spines at more than 7 m (24 ft) are rare, but twin-spine concrete and steel box-girder decks are not uncommon. While the bending and torsional behaviour of the spines must be considered as described below for cellular decks, the distribution of loads between spines is essentially beam-and-slab.

Beam-and-slab decks are most conveniently analysed with the aid of conventional computer grillage programs, as mentioned for slab decks above. The application of this method to these decks is described in Chapter 4. This method is generally accepted as sufficiently accurate for design, but it ignores possible high horizontal shear forces in the slab resulting from differences in the shortening of the top fibres of adjacent beams subjected to different bending deflections. Methods for assessing these forces are described in Chapter 7.

1.2.5 Cellular decks

The cross-section of a cellular or box deck is made up of a number of thin slabs and thin or thick webs which totally enclose a number of cells. These complicated structural forms are increasingly used in preference to beam-and-slab decks for spans in excess of 30 m (100 ft) because in addition to the low material content, low weight and high longitudinal bending stiffness they have high torsional stiffnesses which give them better stability and load distribution characteristics. Nonetheless, some designers consider their popularity a matter of fashion. The use of box decks has been particularly spectacular in recent years for long high spans, where falsework is inappropriate, and the deck has been erected in elements as a beam cantilevering out from supports. Such cantilever construction is less popular with beam-and-slab decks because large trusses are usually needed temporarily to provide torsional stiffness to the incomplete deck. To describe the behaviour of cellular decks it is convenient to divide them into multicellular slabs and box-girders.

Multicellular slabs are wide shallow decks with numerous large cells. The cross-section shape does not lend itself to precast segmental construction, and construction is usually *in situ* concrete or contiguous precast box beams or top hat beams with large voids. For spans up to 36 m (120 ft), the cells of *in situ* concrete decks frequently consist of large cylindrical voids of diameter exceeding 60 per cent of the structural depth. For longer spans such a deck would be too heavy, and the cross-section is lightened by making the cells rectangular and enlarging them so that they fill most of the cross-section with the top and bottom slabs as thin as 200 mm (8 in) and 150 mm (6 in) respectively. When a load is placed on one part of such a deck, the high torsional

Fig. 1.14 Norderelbe Bridge, Hamburg (1962). Cable stayed steel I and box girder deck. Designed and built by Rheinstahl AG, Dortmund. Courtesy of Rheinstahl-Klönne.

stiffness and transverse bending stiffness of the deck transfer and share out the load over a wide area. The distribution is not as effective as that of a slab since the thin top and bottom slabs flex independently when transferring vertical shear forces between webs, and the cross-section is said to 'distort' like a Vierendeel truss in elevation. Such distortion can be reduced by incorporating transverse diaphragms at various points along the deck, but as with beam-and-slab decks their use is becoming less popular except at supports where it is necessary to transfer the vertical shear forces in unsupported webs to neighbouring bearings.

Fig. 1.15 Erskine Bridge, Glasgow (1971). Cable stayed steel box girder deck. Designed by Freeman Fox & Partners, London. Courtesy of Freeman Fox & Partners.

16 *Bridge Deck Behaviour*

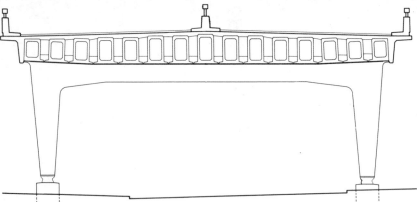

Fig. 1.16   Westway 6, London (1970). Multicellular deck constructed of contiguous top-hat beams. Designed by G. Maunsell & Partners, London. Courtesy of G. Maunsell & Partners (photo); Cement and Concrete Association, London (diagram).

Fig. 1.17 Cell distortion in multicellular decks.

Box-girder decks have a cross-section composed of one or a few large cells, the edge cells often having triangular cross-section with inclined outside web. Frequently the top slab is much wider than the box, with the edges cantilevering out transversely. Excessive twisting of the deck under eccentric loads on the cantilevers is resisted by the high torsional stiffness of the structure. Small and medium span concrete box-girders are usually cast *in situ* or precast in segments which are erected on falsework or launching frame and prestressed. Large spans are generally of segmental cantilever construction with the segments precast or cast *in situ* on a travelling form prior to being stressed back onto the existing structure. Steel box-girders are also frequently constructed as segmented cantilevers, but sometimes they are prefabricated in long lengths, weighing several thousand tons, which are lifted into position by climbing jacks and connected together. The running slabs, of steel battledeck or concrete composite construction, are often added after the torsion boxes are in place.

The method of analysis most appropriate to a particular cellular deck depends on the complexity of the structural form. If the deck has none or few transverse diaphragms a computer shear flexible grillage is adequate, as demonstrated in Chapter 5. In this method the deck is simulated by a grid of beams, as before, but the beams are given the high torsional stiffnesses of the cellular deck, and the slope deflection equations take into account shear deformation in the beams. By attributing very low shear stiffnesses to the transverse beams, their deformation in shear can be made to simulate cell distortion. While generally accurate enough for design, the analysis does not always give a sufficiently detailed picture of the flexural and membrane stresses in the plate elements, and for this additional three-dimensional analyses may be necessary. Space frame analyses, described in Chapter 7, have proved reliable and are liked by several design engineers because the deck is represented by an easily understood physical structure. If the deck has uniform cross-section from end to end and few transverse diaphragms, folded plate analysis (described in Chapter 12) probably provides the most accurate method. On the other hand, if the deck has complicated variations in section and numerous diaphragms, a finite element analysis (described in Chapter 13) may be necessary. This method is complicated and expensive and it is often found most convenient to use it just to study stress

18 Bridge Deck Behaviour

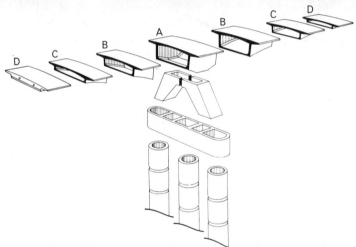

Fig. 1.18 Oosterschelde Bridge, Holland (1965). Designed and built by consortium van Hattum en Blankenvoort N.V. and W. V. Amsterdamsche Ballast Maatschappin. Diagram courtesy of the Cement and Concrete Association, London.

flows in small parts of a structure while simpler methods are used to investigate the load distribution behaviour of the deck as a whole.

REFERENCES

1. Beckett, D. (1969), *Great Buildings of the World: Bridges*, Hamlyn, London.

2
Beam decks : continuous beam analysis

2.1 Introduction

This chapter is concerned with bridge decks which can be thought of as simply supported or continuous beams. After a review of the forms of deck there is a summary of the simple elastic theory of bending and torsion of beams. This summary is intended to be a brief reference to remind the reader of the basic theory on which much of the load distribution analysis of this book is based and which is used for a wide variety of different structures. The study is far from comprehensive, so several references are given to books covering the subject more fully.

2.2 Types of structure

Fig. 2.1a–d shows typical forms of beam decks. The simplest (a), is simply supported on three bearings so that it is statically determinate for bending and torsion. With four bearings in two right pairs as in (b), the deck is statically determinate for bending but not for torsion. Multiple span bridges are often built with spans simply supported as in (c) or continuous as in (d). The statical determinacy in (c) is advantageous when the stability of supports is uncertain due to subsidence or earthquake. However, in general multiple span bridges benefit from having a continuous deck since higher span-to-depth ratios are

20 Bridge Deck Behaviour

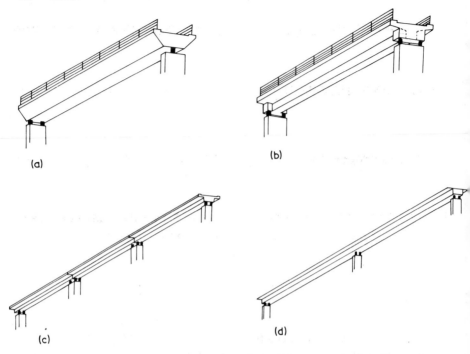

Fig. 2.1 Statical determinacy of bridges. (a) Determinate for bending and torsion (b) determinate for bending only (c) multiple span simply supported determinate for bending (d) multiple span continuous indeterminate.

attainable and the running surface is not interrupted by numerous movement joints. Numerous other articulation and span arrangements are possible as shown in Fig. 2.2. Often the articulation is changed during the construction stages so that it is common for a deck to be statically determinate as beam or cantilever during construction and then made partially or totally continuous for live loads and long term movements.

The bearings of a beam deck can be placed on a skew. If they are also far apart and incompressible, moment and torsion interact significantly at the supports. Analysis of such interaction can be cumbersome by hand, and since it is often not possible with a continuous beam program it may be necessary to use a two-dimensional computer analysis as described in Chapters 4, 5 and 9. In a curved beam deck, moments and torsions interact throughout its length and although hand calculations are possible, the two-dimensional analysis described in Chapter 9 is often more convenient.

It is frequently appropriate to use a continuous beam analysis during the design of a multiple span wide deck with deformable cross-section if the

Beam Decks: Continuous Beam Analysis 21

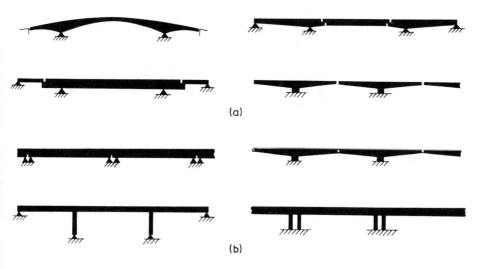

Fig. 2.2 Further examples of determinacy of bridges in bending.
(a) Determinate   (b) indeterminate

supports have little or no skew and the deck is near straight. Under these conditions the distribution of total moment and shear force on cross-sections along the span is virtually the same as that for a single beam with the same total stiffness carrying a load with the same longitudinal disposition. A continuous beam analysis is used to analyse longitudinal distributions of total moments and shear forces due to dead load, construction sequence, prestress, temperature, live load, etc. Then a two-dimensional grillage of about four typical spans is used to determine the transverse distribution of moments, shear forces and accompanying torsion due to lateral disposition of loads and stiffnesses.

2.3 Bending of beams

2.3.1 Equilibrium of forces

Fig. 2.3 shows an element of beam with the right-handed system of axes used throughout the book: Ox is horizontal along the direction of span, Oy is horizontal in the transverse direction, Oz is vertical downwards. Forces in these three directions are denoted by U, V, and W respectively. Deflections are denoted by u, v and w.

The element of a beam in Fig. 2.3 is subjected to a vertical load dW. W can vary along the beam. The element is held in equilibrium by shear forces S and $S + dS$ and moments M and $M + dM$ on the end faces.

22 Bridge Deck Behaviour

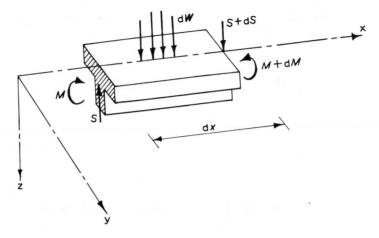

Fig. 2.3   Element of beam.

Resolving vertically we obtain

$$dW = -dS. \tag{2.1}$$

Taking moments about Oy,

$$Sdx = dM$$

or

$$S = \frac{dM}{dx}. \tag{2.2}$$

As shown in Fig. 2.3, positive shear forces 'rotate' clockwise and positive moments cause sagging.

When a simply supported beam supports a load such as the point load in Fig. 2.4a, shear force and bending moment at any point along the beam can be found from a consideration of equilibrium. From these a shear force diagram and bending moment diagram can be drawn as shown in Fig. 2.4b and c. In these diagrams, positive shear force and moment are plotted downwards. Similar diagrams for other common design loads are included in Appendix A.

2.3.2 Stress distributions

In the simple theory of elastic bending of beams it is assumed that the plane sections remain plane and that the beam is composed of discrete linear fibres in which the longitudinal bending stress σ is proportional to the longitudinal strain ϵ in that fibre. From these assumptions it can be shown [1,2] that at any

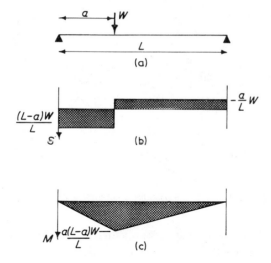

Fig. 2.4   (a) Load   (b) shear force and (c) moment diagram

point on a cross-section σ is proportional to the distance z of the point from the neutral axis, which passes through the centroid of the section. This is illustrated in Fig. 2.5. For flexure about a principal axis of the section

$$\frac{\sigma}{z} = \frac{M}{I} = \frac{E}{R} \qquad (2.3)$$

where M is the total moment on section, I is the second moment of area or

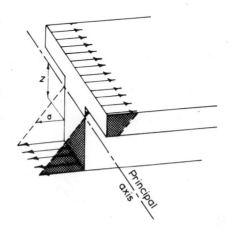

Fig. 2.5   Bending stresses on cross-section.

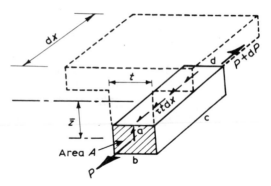

Fig. 2.6 Forces on slice along element of beam.

'moment of inertia' of section about the neutral axis, E is Young's Modulus, and R is the radius of curvature of the beam due to flexure.

If a beam is subjected to loads which are not normal to a principal axis of the section then the loads must be resolved into components normal to the two principal axes. Bending about each axis is then considered separately and the two results superposed.

The shear stress at any point on a cross-section such as a in Fig. 2.6 is found by considering the equilibrium of that part abcd of an element of beam cut off by slice ad along the beam through a. The longitudinal tension force P on end ab is expressed as

P = (average tension stress $\bar{\sigma}$ on ab) × (area A of the face ab).

From Equation 2.3,

$$\bar{\sigma} = \frac{M\bar{z}}{I}$$

where $\bar{z}$ is distance of centroid of area A from centroid of whole section

$$\therefore \quad P = \frac{MA\bar{z}}{I}.$$

Between ends ab and cd of element, P changes by dP:

$$dP = \frac{dMA\bar{z}}{I}.$$

Resolving longitudinally for load on abcd,

$$\tau t dx = dP = \frac{dMA\bar{z}}{I}$$

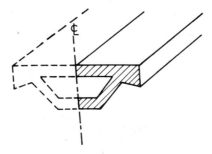

Fig. 2.7 Halving of box deck for analysis of shear stress.

where τ is the longitudinal shear stress along cut ad. Replacing

$\dfrac{dM}{dx}$ by S we obtain

$$\tau t = \dfrac{SA\bar{z}}{I}. \qquad (2.4)$$

Equilibrium of shear stresses at any point requires that they should be complementary in orthogonal directions. Hence Equation 2.4 applies to shear stresses at a in both directions ab and ad.

The distribution of bending shear stress within a box deck such as in Fig. 2.7 can be found with Equation 2.4, but the beam must first be notionally cut longitudinally along its vertical axis of symmetry (on which there are no longitudinal shear stresses). Each side of the deck is then assumed to carry half of the bending shear force with second moment of area I equal to half the total second moment of area of the deck cross-section. If the section is not symmetric, the top and bottom flanges should be cut in the manner described in Section 5.4.1 so that the different sections associated with each web have their individual neutral axes at the same level as the neutral axis of the whole section.

2.3.3 Slopes and deflections

For small deflections

$$R = -\dfrac{1}{\dfrac{d^2 w}{dx^2}}.$$

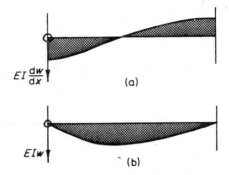

Fig. 2.8 (a) Slope and (b) deflection diagrams for load of Fig. 2.4.

Hence we can write the second half of Equation 2.3 in the form

$$\frac{d^2 w}{dx^2} = -\frac{M}{EI} \tag{2.5}$$

which can be integrated to give

$$\frac{dw}{dx} = -\int \frac{M}{EI} dx \tag{2.6}$$

$$w = -\iint \frac{M}{EI} dx\,dx. \tag{2.7}$$

If E and I are constant along the beam then dw/dx and w can be found by repeated integration of the bending moment diagram. Fig. 2.8a and b shows the diagrams of dw/dx and w for beam of Fig. 2.4, obtained by repeated integration of the bending moment diagram of Fig. 2.4c. Similar slope and deflection diagrams for other common design loads are given in Appendix A. If E and/or I vary along the beam, the bending moment diagram must be divided by EI before integration to obtain dw/dx and w.

2.3.4 Analysis of continuous beams by slope-deflection equations

In contrast to the simply supported beam above, a continuous beam does not generally have zero moments at the supports. Fig. 2.9 shows one span (or part of a span) of a continuous beam subjected to end moments M_{12}, M_{21} and shear forces S_{12} and S_{21} and with end displacements w_1, θ_1 and w_2, θ_2. In Fig. 2.9 and in Equations 2.8 and 2.9 that follow, the end moments are positive clockwise and end shear forces positive downwards. This sign convention is adopted for these equations in preference to the sagging positive convention

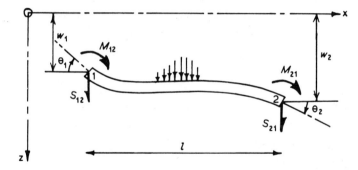

Fig. 2.9 End forces and displacements on part of continuous beam.

above so that moments or shear forces can be added in equilibrium equations without having to consider at which end of the beam they act.

By solving Equations 2.6 and 2.7 for the beam of Fig. 2.9 with its end forces and displacements, the following slope-deflection equations are obtained relating the forces to the deflections:

$$M_{12} = M_{12FE} + \frac{EI}{l}\left[4\theta_1 + 2\theta_2 + \frac{6}{l}(w_1 - w_2)\right]$$

$$M_{21} = M_{21FE} + \frac{EI}{l}\left[2\theta_1 + 4\theta_2 + \frac{6}{l}(w_1 - w_2)\right]$$

$$S_{12} = S_{12FE} + \frac{6EI}{l^2}\left[\theta_1 + \theta_2 + \frac{2}{l}(w_1 - w_2)\right] \quad (2.8)$$

$$S_{21} = S_{21FE} - \frac{6EI}{l^2}\left[\theta_1 + \theta_2 + \frac{2}{l}(w_1 - w_2)\right].$$

Without terms M_{12FE}, M_{21FE}, S_{12FE} and S_{21FE}, Equations 2.8 give the end forces in terms of the end deflections for a beam having no load along its length. If there are intermediate loads acting on the beam, their effect is introduced by the terms M_{12FE}, etc. which are identical to the forces that would act on the end of the beam if the ends were rigidly encastered with $w_1 = w_2 = \theta_1 = \theta_2 = 0$. Fig. 2.10 shows a fixed end beam supporting a point load W at distance x from

Fig. 2.10 Fixed end moments and shear forces.

one end. For this beam the fixed end moments and shear forces are

$$M_{12FE} = -\frac{W(l-x)^2 x}{l^2}$$

$$M_{21FE} = \frac{Wx^2(l-x)}{l^2}$$

(2.9)

$$S_{21FE} = -\frac{W(l-x)^2(l+2x)}{l^3}$$

$$S_{21FE} = -\frac{Wx^2(3l-2x)}{l^3}.$$

The fixed end forces for distributed or varying loads can be obtained by integrating Equations 2.9.

When shear deformation of the beam is significant, Equations 2.8 are modified to [3]:

$$M_{12} = M_{12FE} + \frac{EI}{l}\left[\frac{(1+j)}{(1+4j)}4\theta_1 + \frac{(1-2j)}{(1+4j)}2\theta_2 + \frac{1}{(1+4j)}\frac{6}{l}(w_1 - w_2)\right]$$

$$M_{21} = M_{21FE} + \frac{EI}{l}\left[\frac{(1-2j)}{(1+4j)}2\theta_1 + \frac{(1+j)}{(1+4j)}4\theta_2 + \frac{1}{(1+4j)}\frac{6}{l}(w_1 - w_2)\right]$$

$$S_{12} = S_{12FE} + \frac{6EI}{l^2}\left[\frac{1}{(1+4j)}\theta_1 + \frac{1}{(1+4j)}\theta_2 + \frac{1}{(1+4j)}\frac{2}{l}(w_1 - w_2)\right]$$

$$S_{21} = S_{21FE} - \frac{6EI}{l^2}\left[\frac{1}{(1+4j)}\theta_1 + \frac{1}{(1+4j)}\theta_2 + \frac{1}{(1+4j)}\frac{2}{l}(w_1 - w_2)\right]$$

where

$$j = \frac{3EI}{l^2 GA}$$

$$G = \text{shear modulus} = \frac{E}{2(1+\nu)}$$

A = cross-section area or 'shear area' of section.

Equations 2.8 and 2.9 only apply to beams with constant EI. Similar equations can be written for tapered and haunched beams; useful charts are

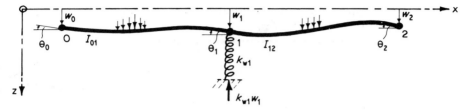

Fig. 2.11 Two spans of continuous beam deck.

contained in reference [3]. However, in a computer analysis it is common for tapered beams to be considered as a number of connected prismatic lengths of reducing size, for each of which Equations 2.8 and 2.9 can be derived.

By considering vertical and rotational equilibrium of any support of a continuous beam such as at 1 in Fig. 2.11, two equations are obtained:

$$a_{01}w_0 + a_{11}w_1 + a_{12}w_2 - b_{01}\theta_0 + b_{11}\theta_1 + b_{12}\theta_2 + S_{10FE} + S_{12FE} = 0$$

$$b_{01}w_0 + b_{11}w_1 - b_{12}w_2 + c_{01}\theta_0 + c_{11}\theta_1 + c_{12}\theta_2 + M_{10FE} + M_{12FE} = 0$$

(2.10)

where

$$a_{01} = -\frac{12EI_{01}}{l_{01}^3} \qquad b_{01} = \frac{6EI_{01}}{l_{01}^2} \qquad c_{01} = \frac{2EI_{01}}{l_{01}}$$

$$a_{11} = -a_{01} - a_{12} + k_{w1} \qquad b_{11} = -b_{01} + b_{12} \qquad c_{11} = 2c_{01} + 2c_{12} + k_{\phi 1}$$

k_{w1} and $k_{\phi 1}$ are vertical and rotational stiffnesses of support 1. Similar pairs of stiffness equations can be written for every other support, so that for N supports there are $2N$ equations for $2N$ unknown deflections. These equations are solved to give the deflections, which can then be substituted back into Equation 2.8 to give the moments and shear forces along the spans. As shown below, the number of equations and unknown deflections can be reduced if supports are rigidly restrained against either vertical or rotational movement.

The above stiffness equations provide a simple method of analysis of continuous beams using programmable desktop calculators. Such machines can usually solve a reasonable number of simultaneous equations. The method is demonstrated below with a worked example. This example can also be solved simply by hand using moment distribution, as described in reference [3].

Worked example
Fig. 2.12a shows a three span bridge with piers 1 and 3 pinned for rotation and rigid vertically, 2 pinned for rotation and with vertical stiffness $k_{w1} = 1000$

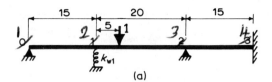

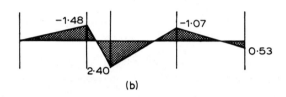

Fig. 2.12 Moments in three span beam.

force/unit deflection, and 4 is rigid against rotation and deflection. $EI = 10\,000$. From Equations 2.9 we find

$$S_{01FE} = S_{10FE} = S_{23FE} = S_{32FE} = M_{01FE} = M_{10FE} = M_{23FE}$$
$$= M_{32FE} = 0$$

$$S_{12FE} = -\frac{1(20-5)^2(20+2\times 5)}{20^3} = -0.844$$

$$S_{21FE} = -\frac{1\times 5^2(3\times 20 - 2\times 5)}{20^3} = -0.156$$

$$M_{12FE} = -\frac{1(20-5)^2 5}{20^2} = -2.81$$

$$M_{21FE} = \frac{1\times 5^2(20-5)}{20^2} = 0.94 .$$

The stiffness coefficients are

$$a_{01} = a_{10} = a_{23} = a_{32} = -\frac{12\times 10\,000}{15^3} = -35.56$$

$$a_{12} = a_{21} = -\frac{12\times 10\,000}{20^3} = -15.0$$

$$a_{11} = 35.56 + 15.0 + 1000 = 1050.56$$

$$b_{01} = b_{10} = b_{23} = b_{32} = \frac{6 \times 10\,000}{15^2} = 266.67$$

$$b_{12} = b_{21} = \frac{6 \times 10\,000}{20^2} = 150.0$$

$b_{00} = 0 + 266.67 = 266.67$

$b_{11} = -266.67 + 150 = -116.67$

$b_{22} = -150 + 266.67 = 116.67$

$$c_{01} = c_{10} = c_{23} = c_{32} = \frac{2 \times 10\,000}{15} = 1333$$

$$c_{12} = c_{21} = \frac{2 \times 10\,000}{20} = 1000$$

$c_{00} = 2 \times 0 + 2 \times 1333 = 2666$

$c_{11} = 2 \times 1333 + 2 \times 1000 = 4666$

$c_{22} = 2 \times 1000 + 2 \times 1333 = 4666.$

Since $w_0 = w_2 = w_3 = \phi_3 = 0$, we do not need to include these in stiffness equations and we can omit equations for corresponding vertical equilibrium at 0, 2 and 3 and rotational equilibrium at 3. Hence writing Equations 2.10 for vertical equilibrium at 1, rotational equilibrium at 0, rotational equilibrium at 1 and rotational equilibrium at 2:

$1050.56 w_1 - 266.67 \theta_0 - 116.67 \theta_1 + 150 \theta_2 \quad\quad -0.844 = 0$

$-266.67 w_1 + 2666 \theta_0 + 1333 \theta_1 \quad\quad\quad\quad\quad\quad\quad\quad = 0$

$-116.67 w_1 + 1333 \theta_0 + 4666 \theta_1 + 1000 \theta_2 \quad -2.81 = 0$

$150 \quad w_1 \quad\quad\quad\quad + 1000 \theta_1 + 4666 \theta_2 + 0.94 \quad\quad = 0$

which can be solved to give

$w_1 = 0.000\,87 \quad\quad \theta_0 = -0.000\,313 \quad\quad \theta_1 = 0.000\,799 \quad\quad \theta_2 = -0.0004\,0$

By substituting these back into Equations 2.8 we obtain the moment diagram of Fig. 2.12b.

2.3.5 Analysis of continuous beams by flexibility coefficients

The preceding stiffness equations were derived by assuming the structure has certain unknown deflections at supports and then deriving equilibrium equations for each support in turn in terms of the deflections. The equation for equilibrium of moments or vertical forces at any support can be thought of as

$$\Sigma[(\text{force on joint 1 due to unit deflection at 2}) \times (\text{deflection at 2})] + \text{applied load on 1} = 0.$$

where Σ is sum for all deflections 1 to n

or

$$\Sigma[\text{stiffness coefficient} \times \text{deflection}] + \text{applied load} = 0.$$

An alternative approach, using 'flexibility' or 'influence' coefficients is sometimes more convenient and is demonstrated below. The indeterminate structure such as the continuous beam in Fig. 2.13a is notionally 'released' in a number of places so that it becomes statically determinate. The problem is then to determine what equal and opposite actions X are to be applied to the two sides of every release so that all the releases close. To calculate the values of the indeterminate actions X, the opening of each release is calculated due to unit values of each action X in turn and due to the applied load. Then for release at 1 to close, for example, the sum of the openings of the release due to all the loads is zero, which expressed as an equation is

$$\Sigma[(\text{opening of release 1 due to unit action } X_2) \times X_2] + \text{opening of release 1 due to applied load} = 0$$

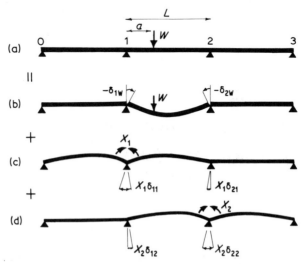

Fig. 2.13 Continuous beam = superposition of released beams.

where Σ is the sum for all actions 1 to n

or

Σ[flexibility coefficient × action X] + opening due to applied load = 0.

By writing similar flexibility equations for all of the releases and solving for the actions X we obtain the values of the actions relevant for all the cuts to close simultaneously.

The above equations are expressed algebraically by

$$\begin{aligned}\delta_{11}X_1 + \delta_{12}X_2 + \ldots \delta_{1n}X_n + \delta_{1W} &= 0 \\ \delta_{21}X_1 + \delta_{22}X_2 + \ldots \delta_{2n}X_n + \delta_{2W} &= 0 \\ \delta_{n1}X_1 + \delta_{n2}X_2 + \ldots \delta_{nn}X_n + \delta_{nW} &= 0\end{aligned} \quad (2.11)$$

where δ_{jk} = opening of cut j due to unit action X_k,

δ_{jW} = opening of cut j due to applied loads.

From considerations of flexural strain energy [4] it can be shown

$$\delta_{12} = \int \frac{m_1 m_2}{EI} dx$$

$$\delta_{1W} = \int \frac{m_1 m_W}{EI} dx \quad (2.12)$$

where, as illustrated in Figs. 2.14, m_1 and m_2 are moments in released structure due to actions $X_1 = 1$ and $X_2 = 1$ respectively, and m_W are the moments in the released structure due to the applied loads. The values of the product integrals $\int mm dx$ for common moment diagrams are given in Appendix A, Fig. A.1. Their application is demonstrated below.

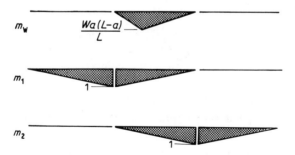

Fig. 2.14 Moments due to applied loads and unit release actions in Fig. 2.13.

Worked example

Fig. 2.15 shows a four span bridge deck supporting a point load near the middle of the second span. Fig. 2.15b shows the applied load moment diagram if the structure is released by relaxing the moments at 1, 2 and 3. Figs. 2.15c–e show the moment diagrams m_1, m_2 and m_3 due to unit actions at 1, 2 and 3 respectively. From Appendix A we obtain:

$$\delta_{11} = \left(\frac{10}{3} + \frac{15}{3}\right)\frac{1}{EI} = \frac{8.33}{EI} \qquad \delta_{12} = \left(\frac{15}{6}\right)\frac{1}{EI} = \frac{2.5}{EI} \qquad \delta_{13} = 0$$

$$\delta_{22} = \left(\frac{15}{3} + \frac{15}{3}\right)\frac{1}{EI} = \frac{10}{EI} \qquad \delta_{23} = \left(\frac{15}{6}\right)\frac{1}{EI} = \frac{2.5}{EI}$$

$$\delta_{33} = \left(\frac{15}{3} + \frac{10}{3}\right)\frac{1}{EI} = \frac{8.33}{EI}$$

$$\delta_{1W} = \frac{15}{6} \times 3.33 \times 1 \times (1 + 0.66) \times \frac{1}{EI} = \frac{13.87}{EI}$$

$$\delta_{2W} = \frac{15}{6} \times 3.33 \times 1 \times (1 + 0.33) \times \frac{1}{EI} = \frac{11.1}{EI}.$$

Hence the flexibility equations can be written, (omitting $1/EI$)

$$8.33X_1 + 2.5X_2 \qquad\qquad + 13.87 = 0$$
$$2.5\ X_1 + 10X_2 + 2.5X_3 + 11.10 = 0$$
$$\qquad\quad 2.5X_2 + 8.33X_3 + 0 \quad\; = 0$$

which when solved give the unknown moments over the supports

$$X_1 = -1.42 \quad X_2 = -0.82 \quad X_3 = +0.24.$$

These are shown in Fig. 2.15f.

It should be noted that relaxation of the moments at the supports is not the only way the structure could have been released. It could be done by removing the supports at 1, 2 and 3 so that the deck spans simply supported from 0 to 4. Then actions X_1, X_2 and X_3 would be vertical forces and would be calculated to produce zero vertical deflection at the release points (or deflections equal to support flexibility × vertical reactions).

The selection of the flexibility method or the stiffness method for a particular problem depends on the form of the structure, its supports, and loading, and also on the designer as to whether he finds it easier to 'feel' the stiffness of a member or the effect of a release displacement. In general the

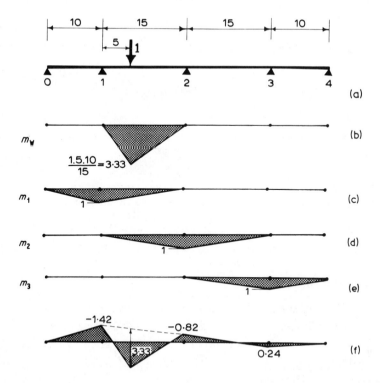

Fig. 2.15 Flexibility analysis of continuous beam. (a) Load (b) load moments (c–e) unit release moments (f) final moment diagram.

stiffness method is more popular, and easier to use in computer programs. Both methods are described in greater detail in reference [5].

2.3.6 Multispan beams

During the analysis of beams with many approximately equal spans of uniform section it is often convenient to reduce the amount of calculation by notionally replacing the unloaded spans to the side of the region of consideration by a single span of approximately 0.87 times its real length, as shown in Fig. 2.16b. The moment calculated for the penultimate support in the reduced structure will be close to that in the extended structure. Having determined this moment, those in the omitted internal supports are found simply by progressing from the loaded region and making each support moment -0.27 of the preceding support moment as shown in Fig. 2.16c.

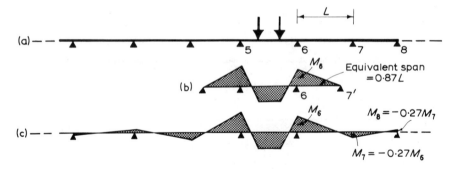

Fig. 2.16 Approximate analysis of multiple span beams. (a) Prototype (b) approximate analysis of reduced structure   (c) moments in load free regions.

2.4 Torsion of beams

2.4.1 Equilibrium of torques

In Fig. 2.3 the vertical load dW was acting above the centre line of the beam and was held in equilibrium solely by the shear forces S and moments M. If the vertical load dW is placed eccentric to the centre line as in Fig. 2.17 then additional actions in the form of torques T and $T + dT$ on the ends of the element are necessary to retain equilibrium of couples about the longitudinal axis Ox. On taking moments about Ox we obtain

$$dT = y\,dW \tag{2.13}$$

where y is the eccentricity of the load from the longitudinal centre line. If the beam does not have a vertical axis of symmetry then y should be measured from the shear centre [1,2]. Since most beam decks are symmetrical, this added complexity is not dealt with further here.

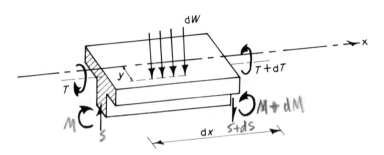

Fig. 2.17   Torque on element of beam.

Beam Decks: Continuous Beam Analysis   37

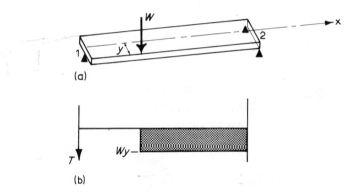

Fig. 2.18 Torque diagram. (a) Beam subjected to eccentric load   (b) torque diagram.

Fig. 2.18a shows a simply supported beam deck, with two bearings at end 2 and one bearing at end 1. Under the action of the eccentric load W near midspan the beam is subjected to a torque. With only one bearing at end 1, no torque can be transmitted to the support. To determine the torque T on any cross-section, moments can be taken about Ox for the beam to the left of the section. In this way, the torque diagram in Fig. 2.18b is obtained.

It should be noticed that a beam is only statically determinate for torsion when it has only one pair of bearings to resist torques. Fig. 2.19 shows a number of different bearing arrangements for beams which are: statically determinate in both bending and torsion, statically determinate in torsion but not bending, statically determinate in bending but not torsion, and statically indeterminate in

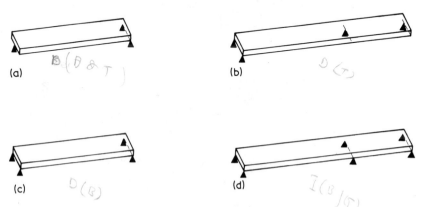

Fig. 2.19 Statical determinacy of beam in torsion. (a) Determinate for bending and torsion   (b) determinate for torsion only   (c) determinate for bending only   (d) indeterminate for both bending and torsion.

38 Bridge Deck Behaviour

both bending and torsion. If all pairs of bearings are placed at right angles to the longitudinal axis then the equilibrium (and analysis) of the beam in bending and torsion can be considered separately. However if a pair of bearings are skew at a pier, the moments and torques interact at the pier and the analysis is more complicated and is best performed with a two-dimensional grillage as described in Chapter 4. An exception is the simple case of a beam supported on only three bearings, not in line, when all the reactions can always be determined by taking moments about two axes.

2.4.2 Torque-twist relationship

To determine the distribution of torques in a beam with four bearings, as in Fig. 2.20, some knowledge of its deformation behaviour is needed. It is only if the beam has uniform torsional stiffness that simple physical argument leads to the solution.

When a torque T is applied to an element of beam it causes the element to twist about axis Ox with a relative rotation $d\phi$ between the ends of the element. For an elastic material, the amount of relative rotation is proportional to the torque and related by the equation

$$T = -CG\frac{d\phi}{dx} \tag{2.14}$$

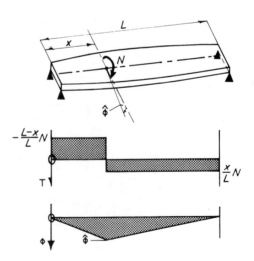

Fig. 2.20 Torsion of beam with both ends restrained. (a) Loading (b) torque diagram (c) rotation diagram.

where C is the torsion constant of the section, discussed below. G is the Shear Modulus = $E/2(1 + \nu)$ where ν is Poisson's Ratio

$$d\phi = -\frac{T}{CG} dx$$

$$\phi = -\int \frac{T}{CG} dx. \qquad (2.15)$$

The distribution of torque and rotation along a beam of uniform torsional rigidity CG restrained at both ends against rotation can be found as follows. Fig. 2.20 shows such a beam supporting a concentrated applied torque N at some point along the span. Since there are no other loads between N and the ends the torques must be constant (though different) along these lengths, as shown in Fig. 2.20b. The rotation ϕ varies along the beam from zero at a support to $\hat{\phi}$ at the load and back to zero at the other support. Since the torques are constant on each side of the load so must $d\phi/dx$ be, from Equation 2.14. Hence the ϕ diagram in Fig. 2.20c must be triangular, and $d\phi/dx$ on each side must be inversely proportional to the distance of the load from each end. Consequently the applied torque N is distributed to the two supports in fractions inversely proportional to the distance of the load from the support. It is thus found for this concentrated applied torque that the distribution of torque is similar to the distribution of shear force due to concentrated vertical load and the distribution of ϕ similar to the distribution of moment. Since a distributed load can be thought of as a superposition of concentrated loads, it is possible to derive the distribution of T and ϕ for any load on a uniform beam between two torque resisting supports by direct analogy with shear force and moment distributions for a simple span. It should be noted that this analogy does not hold if CG is not constant.

2.4.3 Torsion constant C

The torsion constant C (often referred to as the Saint-Venant torsion constant) is not generally a simple geometrical property of the cross-section in the same way that the flexural constant I is the second moment of area. In the case of a cylinder, C is identical to the polar moment of inertia I_p. However this is a special case which can be misleading, since for many shapes of cross-sections C is totally different from I_p, and can differ by an order of magnitude. There is not a general rule for the derivation of C or for the analysis of torsional shear stress distribution. References [2 and 6] outline the elastic theory of torsion of prismatic beams for a number of shapes of cross-section. The following

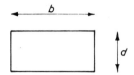

Fig. 2.21 Solid rectangular cross-section.

paragraphs give approximate rules appropriate to the shapes most common to beam bridges.

C of solid cross-sections without reentrant corners
Saint-Venant derived an approximate expression which is applicable to all cross-section shapes without reentrant corners, i.e. triangles, rectangles, circles, ellipses, etc. The expression is

$$C = \frac{A^4}{40 I_p} \qquad (2.16)$$

where A = area of cross-section and I_p = polar moment of inertia.

For a rectangle of sides b and d shown in Fig. 2.21, this can be reduced to

$$C = \frac{3 b^3 d^3}{10(b^2 + d^2)} \qquad (2.17)$$

and in the case of a thin rectangle $b > 5d$ this is more accurately given by

$$C = \frac{b d^3}{3}. \qquad (2.18)$$

The maximum shear stress on a rectangular cross-section occurs at the

Fig. 2.22 Subdivision of section with reentrant corners.

Fig. 2.23   Inflated membrane on cut rectangular cross-section.

middle of the long side and has magnitude

$$\hat{\tau} = \frac{T}{bd^2(0.333 - 0.125\sqrt{d/b})} \qquad b > d. \qquad (2.19)$$

C of solid cross-sections with reentrant corners
If the cross-section has reentrant corners, C is very much less than that given by Equation 2.16. C is obtained by notionally subdividing the cross-section, as in Fig. 2.22, into shapes without reentrant corners and summing the values of C for these elements. While doing this it is worth remembering Prandtl's membrane analogy, described in reference [6]. It is shown that the stiffness of a cross-section shape is proportional to the volume under an inflated bubble stretched across a hole of the same shape. The shear stress at any point is along the direction of the bubble's contours and of magnitude proportional to the gradient at right angles to the contours. If a cross-section is cut in half, the membrane is in effect held down along the cut as in Fig. 2.23, thus greatly reducing its volume and preventing flow of shear stress along the contours from one part to the other. Consequently when a section such as Fig. 2.22 is notionally split into elements it is important to choose the elements so that they maximize the volume under their bubbles. To avoid giving the bubble zero height at the notional cuts, the elements can be rejoined at the cuts for calculation as in Fig. 2.22b. Since shear stresses flow across both ends of the web it can be thought of as part of a long thin rectangle for which $C = bd^3/3$. Fig. 2.24 also shows the cross-section arbitrarily cut into rectangles which, by not trying to maximize the volume under the bubble, leads to a value of C of only half the correct figure.

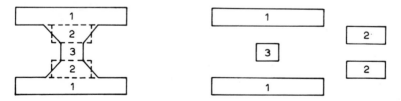

Fig. 2.24   Erroneous subdivision of section with reentrant corners.

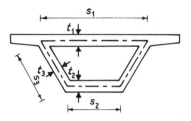

Fig. 2.25   Thin-walled box section.

C of thick-walled hollow sections

The torsion constant of a thick-walled hollow section can be found with sufficient accuracy for most practical purposes by calculating C for the shape of the outside boundary and deducting the value of C calculated for the inside boundary.

C of thin-walled hollow sections

The torsion constant of a thin-walled hollow section, such as in Fig. 2.25, is given by

$$C = \frac{4A^2}{\oint \frac{ds}{t}} \tag{2.20}$$

where A is the area enclosed by the centre line of the walls and $\oint ds/t$ is the integral round the wall centre line of the length divided by the wall thickness.

The shear stress round the wall at any point is given by

$$\tau = \frac{T}{2\,At} \tag{2.21}$$

Equation 2.20 is only really applicable for cross-sections with only one cell or two symmetric cells. The equation can be applied to the outside boundary of a multicellular box as a first approximation, but for a more accurate calculation see references [2 and 6].

2.5 Computer analysis of continuous beams

There are numerous different computer programs for analysis of continuous bridges. In general they are simple to use, and the derivation of computer input section properties is straightforward. It is not possible to mention all the facilities that are available, but convenient programs exist for the analysis of beams which are simply supported, continuous, prismatic, varying in section, supported or restrained by elastic supports. The analysis of a large number of load cases including prestress, settlement and temperature can be requested by few instructions. Output can be in terms of bending moment and shear force distributions, displacements, influence lines, or envelopes of maximum and minimum moments and shear forces along the deck. Not all these facilities are necessarily available in a single program; in fact as a general rule, the more versatile a program is, the more difficult it is to use. It is thus worth choosing the program to suit the complexity of the problem, unless a user has sufficient familiarity with one program in particular to enable him to adapt his use of it as necessary.

2.6 Construction sequence

The analysis of live load actions is generally a straightforward procedure of applying a number of different load cases with various dispositions and intensities to the final structure, and inspecting moments and shear forces at critical design sections. With experience, the critical load cases are quickly recognized. In contrast the analysis of dead load and prestress of a continuous deck built in a number of different stages can be complex. Unless the design is very conservative, the method of construction must be considered during design and, vice versa, the method of design must be remembered during construction.

If a structure is built from end to end on falsework prior to removal of the falsework, then the dead load moments should be as calculated for simultaneous application of the dead load to continuous beam, as shown in Fig. 2.26. On the other hand, if the bridge is erected span by span, as were the boxes of Stephenson's Britannia Bridge, simple moment connections of the spans over the supports do not of themselves induce large dead load moments over the supports which reduce span moments. These moments can only be induced by removing the relative rotations due to self weight of the touching ends of adjacent spans. This can be done by jacking a rotation at the joint, or by bolting up with prestressing bolts, or as Stephenson did by connecting spans as in Fig. 2.27 with one end temporarily raised. The articulation and moments of his boxes changed

Fig. 2.26 Moments in continuous beam due to simultaneous application of dead load from end to end.

Fig. 2.27 Induction of internal support moments by lowering ends.

as each span was added, and each had to be designed as simply supported for dead load during construction and continuous for dead load and live load in use.

Multiple span cast *in situ* concrete bridges are often built span by span or two spans at a time with cantilevers as in Fig. 2.28. The simplest way of visualizing the dead loading due to any stage of construction is to consider the structure just before and after the falsework is struck. The load added to the continuous deck of cured concrete is identical to the load previously carried by the falsework. The bending moment diagram in Fig. 2.28 with alternate high and low moments at supports is typical of 2-span by 2-span construction. If the deck is prestressed in the same stages, the falsework is effectively struck when the prestress picks up the dead load (except near supports where falsework will carry part of the reaction until its elastic compression is released). If the prestress profile is the same in every span it also generates supports moments of alternating magnitude which, being opposite in sign to dead load moments,

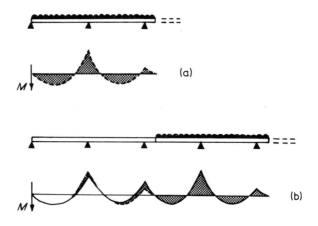

Fig. 2.28 Two span by two span construction.

virtually cancel out the effects of stage by stage construction. The analysis of prestress moments is discussed in Chapter 11.

2.7 Redistribution of moments

Most construction materials creep over a period of time under continuous application of dead load and prestress. This behaviour can be visualized by thinking of the deck as made of putty. After a period of time any peculiarities of deflected shape and curvature due to construction sequence are masked by the deflections of the complete structure due to creep. The distributions of moments and shear forces after long term creep tend towards the distributions that would exist in the structure if it had been built simultaneously from end to end on falsework. This is because introduction of viscous terms into beam Equations 2.5, etc. only has the effect of changing all deflection terms to rates of change of deflection and does not alter their distributions. Hence an indication of the ultimate effects of creep on a multiple span deck built in stages can be obtained by carrying out a 'simultaneous construction' analysis in addition to the stage by stage construction analysis.

2.8 Beam decks with raking piers

Several continuous beam decks have been built with raking piers. The action of a vertical load on a deck induces bending and compression in the deck beam and sway displacements. The combined bending and axial load stresses can differ significantly from those obtained from a continuous beam program so that a plane frame analysis is necessary.

REFERENCES

1. Case, J. and Chilver, A. H. (1959), *Strength of Materials*, Edward Arnold, London.
2. Oden, J. T. (1967), *Mechanics of Elastic Structures*, McGraw-Hill, New York.
3. Lightfoot, E. (1961), *Moment Distribution*, E. & F. N. Spon, London.
4. Morice, P. B. (1959), *Linear Structural Analysis*, Thames and Hudson, London.
5. Coates, R. C., Coutie, M. G. and Kong, F. K. (1972), *Structural Analysis*, Nelson, London.
6. Timoshenko, S. and Goodier, J. N. (1951), *Theory of Elasticity*, McGraw-Hill, New York.

3
Slab decks : grillage analysis

3.1 Introduction

A slab deck is structurally continuous in the two dimensions of the plane of the slab so that an applied load is supported by two-dimensional distributions of shear forces, moments and torques. These distributions are considerably more complex than those along a one-dimensional continuous beam. This chapter presents the fundamental relationships of equilibrium and stress-strain behaviour of an element of a slab. Because rigorous solution of the basic equations for a real deck is seldom possible, a much used approximate method is described. This is grillage analysis in which the deck is represented for the purpose of analysis by a two-dimensional grillage of beams. Another approximate method called 'finite element analysis' is described in Chapter 13. In this method the deck is notionally subdivided into a large number of small elements for each of which approximate plate bending equations can be written and the whole set solved. Hand methods of analysis employing charts are also frequently used for slab decks with simple plan geometry. These methods are reviewed in Chapter 10. However, the steady improvement made to grillage programs in recent years now make this computer method more versatile, as quick, and simpler to comprehend than chart methods.

3.2 Types of structures

Fig. 3.1 shows some common forms of slab deck construction. In Fig. 3.1a the slab is of solid reinforced concrete. In (b) the weight has been reduced by casting voids within the depth of the slab, and the deck is referred to as a 'voided slab'. If the depth of the voids exceeds 60 per cent of the depth of the slab, the slab may not behave like a single plate but more like a cellular deck for which analysis is described in Chapter 5. A slab deck can be constructed of composite construction as in Fig. 3.1c and d. In (c) the slab has been constructed by casting in-fill concrete between contiguous beams with continuous transverse reinforcement top and bottom. In (d) the deck is constructed of contiguous box beams post-tensioned transversely to give moment continuity.

The slab decks in Fig. 3.1 can have similar stiffnesses in longitudinal and transverse directions in which case they are called 'isotropic'. If the stiffnesses

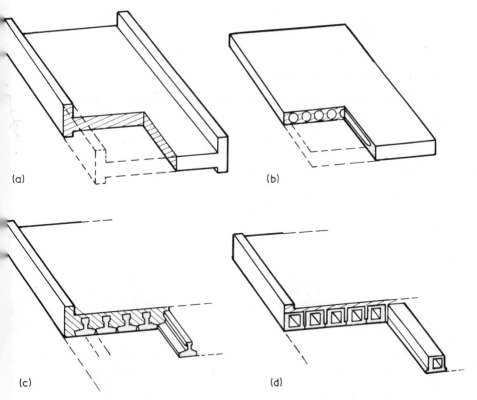

Fig. 3.1 Slab decks. (a) Solid (b) voided (c) composite solid (d) composite voided.

48   Bridge Deck Behaviour

differ in the two directions, as is likely for the decks of (c) and (d), then the slab is called 'orthotropic'.

Slab decks sometimes have their parapet edges stiffened by upstand and downstand beams as in Fig. 3.1a. The deck is then equally able to carry a load in the centre with load distributed to the slab on both sides, and to carry the load near one edge with distribution to the slab on one side and stiffening beam on the other. While such edge stiffening presents problems for rigorous analysis, it does not complicate the approximate methods described later unless the upstand or downstand is so deep that the neutral axis locally is at a significantly different level from the mid-plane of the slab. The effects of such upstand beams and of service troughs are discussed in Chapter 8.

Bridges are frequently designed with their decks skew to the supports, tapered, or curved in plan. The behaviour and rigorous analysis is significantly complicated by the shape, but as shown in Chapter 9 the effect on grillage analysis is one of inconvenience rather than theoretical complexity.

3.3 Structural action

3.3.1 Equilibrium of forces

Fig. 3.2 shows an element of slab subjected to vertical load dW and internal moments m, shear forces s and torques t (all per unit width) which interact with adjoining parts of the slab. By writing

$$\frac{\partial m_y}{\partial y} dy = dm_y, \frac{\partial s_y}{\partial y} dy = ds_y \text{ etc, and } Wdxdy = dW$$

we obtain on resolving vertically and taking moments about axes Ox and Oy, after simplification

$$\frac{\partial s_x}{\partial x} + \frac{\partial s_y}{\partial y} = -W \tag{3.1}$$

$$\frac{\partial m_x}{\partial x} + \frac{\partial t_{yx}}{\partial y} = s_x$$

$$\frac{\partial m_y}{\partial y} + \frac{\partial t_{xy}}{\partial x} = s_y. \tag{3.2}$$

These equations differ significantly from those for a single beam. In addition to the obvious difference that loads distribute in two dimensions, Equations 3.2

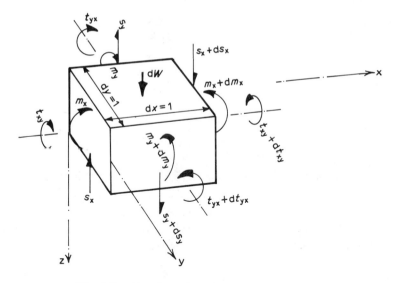

Fig. 3.2 Resultant forces on element of slab.

indicate that shear force is not the simple differential of the bending moment (i.e. it is not the slope of the bending moment diagram).

In grillage analysis the different components $\partial m_x/\partial x$ due to bending and $\partial t_{yx}/\partial y$ due to torsion exhibit themselves in different ways, and it is convenient to define

$$s_{Mx} = \frac{\partial m_x}{\partial x} \quad \text{and} \quad s_{Tx} \frac{\partial t_{yx}}{\partial y} \tag{3.3}$$

so that Equation 3.2 becomes

$$s_x = s_{Mx} + s_{Tx}.$$

At any level of the slab element the horizontal shear stresses on faces normal to Ox and Oy must be complementary to maintain equilibrium. Consequently, the torques on orthogonal faces of the slab element are also complementary and equal:

$$t_{xy} = t_{yx}. \tag{3.4}$$

3.3.2 Moment-curvature equations

The simple theory of elastic bending of slabs is based on the same assumptions as simple beam theory. Lines in the slab normal to the neutral plane remain straight

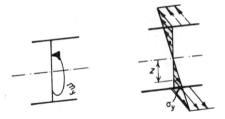

Fig. 3.3 Bending stress distribution.

so that strains and bending stresses shown in Fig. 3.3 increase linearly with distance from the neutral axis. Also, vertical compressive stresses are zero. However, unlike a simple beam, the compressive bending stress σ in one direction is dependent on the compressive strain in the orthogonal direction as well as the compressive strain in its own direction, i.e.

$$\frac{\sigma_x}{z} = \frac{m_x}{i} = \frac{E}{(1-\nu^2)}\left(\frac{1}{R_x} + \frac{\nu}{R_y}\right)$$

$$\frac{\sigma_y}{z} = \frac{m_y}{i} = \frac{E}{(1-\nu^2)}\left(\frac{1}{R_y} + \frac{\nu}{R_x}\right) \tag{3.5}$$

or

$$m_x = -\frac{Ed^3}{(1-\nu^2)12}\left(\frac{\partial^2 w}{\partial x^2} + \nu\frac{\partial^2 w}{\partial y^2}\right)$$

$$m_y = -\frac{Ed^3}{(1-\nu^2)12}\left(\frac{\partial^2 w}{\partial y^2} + \nu\frac{\partial^2 w}{\partial x^2}\right)$$

where

z = vertical distance of point below neutral axis

$i = d^3/12$ = second moment of area of slab per unit width

d = slab thickness

R_x = radius of bending curvature in the x direction

E = Young's Modulus

ν = Poisson's Ratio.

The torsion shear stresses in an element of slab have a linear distribution as shown in Fig. 3.4 with stress τ proportional to distance z from the neutral axis,

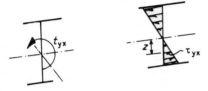

Fig. 3.4   Torsion stress distribution.

so that

$$\frac{\tau_{xy}}{z} = \frac{t_{xy}}{i} = -\frac{E}{(1+\nu)}\left(\frac{1}{R_{xy}}\right)$$

$$t_{xy} = -\frac{Ed^3}{(1+\nu)12}\left(\frac{\partial^2 w}{\partial x \partial y}\right) = -\frac{Gd^3}{6}\left(\frac{\partial^2 w}{\partial x \partial y}\right) \quad (3.6)$$

where

$$G = \frac{E}{2(1+\nu)} = \text{Elastic Shear Modulus.}$$

Equation 3.6 for t_{xy} can be written

$$t_{xy} = -cG\frac{\partial^2 w}{\partial x \partial y} \quad (3.7)$$

where c is the effective torsion constant per unit width of slab given by

$$c = \frac{d^3}{6} \text{ per unit width.} \quad (3.8)$$

Equation 3.8 for the torsion constant of a slab per unit width is equal to half that in Equation 2.18 for a thin slab-like beam. This difference is the consequence of a difference in definition of torque. If the twisted thin slab-like beam in Fig. 3.5 is analysed as a beam as per Section 2.4, then the torque T is defined as the sum of the torque due to the opposed horizontal shear flows near the top and bottom faces and of the torque due to the opposed vertical shear flows near the two edges. In contrast, if the slab-like beam of Fig. 3.5 is analysed as a slab, then the torque t_{xy} is defined as only due to the opposed horizontal shear flows near the top and bottom faces. The vertical shear flows at the edges constitute local high values of the vertical shear force s_x. The opposed vertical shear flows provide half the total torque and are associated by Equation 3.3 with the transverse torque t_{yx} defined in Fig. 3.2. The two definitions of torque, though different, are equivalent: while the slab has half the torsion constant (and hence half strain energy) of the 'beam' attributed to longitudinal torsion,

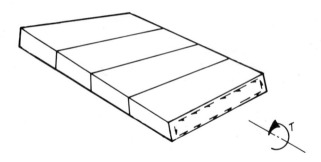

Fig. 3.5   Torsion of slab-like beam.

the other half of the torsion constant (and strain energy) is attributed to transverse torsion not considered in beam analysis.

Equations 3.5–3.8 relate to isotropic slabs whose elastic behaviour can be described by the constants E and ν. If the slab is orthotropic, Young's modulus and Poisson's Ratio are different in the two directions and the moment curvature equations are much more complicated.

$$m_x = -D_x \left(\frac{\partial^2 w}{\partial x^2} + \nu_y \frac{\partial^2 w}{\partial y^2} \right)$$

$$m_y = -D_y \left(\frac{\partial^2 w}{\partial y^2} + \nu_x \frac{\partial^2 w}{\partial x^2} \right)$$

$$t_{xy} = -2D_{xy} \left(\frac{\partial^2 w}{\partial x \partial y} \right)$$

where (3.9)

$$D_x = \frac{E_x d^3}{(1-\nu_x \nu_y)12}$$

$$D_y = \frac{E_y d^3}{(1-\nu_x \nu_y)12}$$

$$D_{xy} = \frac{E_x E_y d^3}{[E_x(1+\nu_{yx}) + E_y(1+\nu_{xy})]12} = \frac{G_{xy} d^3}{12}$$

$$\simeq \tfrac{1}{2}(1-\nu_x \nu_y)\sqrt{(D_x D_y)}.$$

The expression for D_{xy} is an approximation determined by Huber [1].

3.3.3 Principal bending moments and principal stresses

The element of slab in Fig. 3.6a has been defined with faces normal to axes Ox and Oy. The faces are subjected to combinations of moment and torsion $m_x, t_{xy}; m_y, t_{yx}$. If the element is defined with faces normal to axes in other directions, the magnitudes of the moments and torques are different. With axes in one particular set of directions called the principal directions, the torques disappear as in Fig. 3.6b and the moments m_I and m_{II} on the faces represent the maximum and minimum moments at that point in the slab. If α is the angle between the axis Ox in Fig. 3.6a and the axis I-I in Fig. 3.6b, the principal moments m_I and m_{II} are related to m_x, m_y and t_{xy} by the equation

$$m_I = \frac{m_x + m_y}{2} + \sqrt{\left[\left(\frac{m_x - m_y}{2}\right)^2 + t_{xy}^2\right]}$$

$$m_{II} = \frac{m_x + m_y}{2} - \sqrt{\left[\left(\frac{m_x - m_y}{2}\right)^2 + t_{xy}^2\right]}$$

(3.10)

$$\tan 2\alpha = \frac{2t_{xy}}{m_x - m_y}$$

This relationship between moments and torques on faces normal to various axes can be represented by Mohr's circle shown in Fig. 3.6c. With axes for moment m and torque t (here defined as positive for right-handed screw torque vectors towards the centre of the element, so that $t_{yx} = -t_{xy}$) the circle is drawn with points (m_x, t_{xy}) and (m_y, t_{yx}) at opposite ends of a diameter. The points $(m_I, 0)$ and $(m_{II}, 0)$ also on opposite ends of a diameter represent the major and minor principal moments. It should be noted that a difference in direction of axes of α in the element is represented by a difference in inclination of diameters of 2α on Mohr's circle. The maximum torque occurs on the circle at

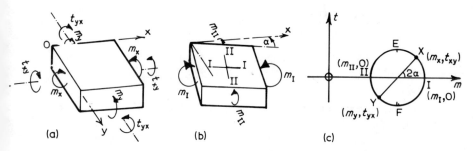

Fig. 3.6 Principal moments and Mohr's Circle of moment.

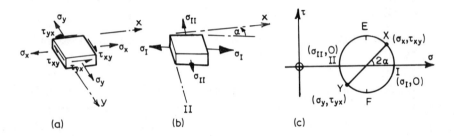

Fig. 3.7 Principal stresses and Mohr's Circle of stress.

points E and F at which moments are equal and which are at opposite ends of the diameter at 90° to $m_I m_{II}$. In the element, the maximum torques occur on faces normal to axes at 45° to principal axes. From the geometry of the circle it is evident that the maximum torque at E or F is

$$\hat{t} = \frac{m_I - m_{II}}{2} = \sqrt{\left[\left(\frac{m_x - m_y}{2}\right)^2 + t_{xy}^2\right]}. \tag{3.11}$$

Tensile stresses and shear stresses on orthogonal planes through a point in the slab, as in Fig. 3.7, are related by the same rules of equilibrium as moments and torques so that Mohr's circle is also used. The sign convention for shear stresses is positive stresses are in a direction clockwise round the element, so that $\tau_{yx} = -\tau_{xy}$. Consequently the compressive and shear stresses σ_x, σ_y, τ_{xy} on faces normal to axes Ox and Oy and the principal stresses σ_I, σ_{II} with axes inclined at α to Oxy are related by

$$\sigma_I = \frac{\sigma_x + \sigma_y}{2} + \sqrt{\left[\left(\frac{\sigma_x - \sigma_y}{2}\right)^2 + \tau_{xy}^2\right]}$$

$$\sigma_{II} = \frac{\sigma_x + \sigma_y}{2} - \sqrt{\left[\left(\frac{\sigma_x - \sigma_y}{2}\right)^2 + \tau_{xy}^2\right]} \tag{3.12}$$

$$\tan 2\alpha = \frac{2\tau_{xy}}{\sigma_x - \sigma_y}.$$

The maximum shear stress acts on planes with axes at 45° to principal axes and is given by

$$\hat{\tau} = \frac{\sigma_I - \sigma_{II}}{2} = \sqrt{\left[\left(\frac{\sigma_x - \sigma_y}{2}\right)^2 + \tau_{xy}^2\right]}. \tag{3.13}$$

3.4 Rigorous analysis of distribution of forces

Manipulation of Equations 3.1 — 3.9 for the analysis of distribution of moments, etc. throughout a slab is complex. Rigorous solutions have been obtained for a few simple shapes of plate under particular load distributions [3—5], but no generally applicable method of rigorous solution has been found. Furthermore, no bridge deck rigorously satisfies the assumptions of isotropic or orthotropic behaviour with the result that assumptions of simplified structural action are necessary to interpret structural details into mathematical stiffnesses. Thus it is in general both impossible to develop rigorous mathematical equations to represent a structure and also impossible to solve the equations once obtained. However, approximate methods are available which either solve the plate bending equations by approximate numerical methods, or which represent the two-dimensional continuum of the deck by an assemblage of small elements or grillage of beams. The latter methods, which have only been practical since the advent of the computer, have the advantages of direct physical significance to engineers and versatility in representing different stiffnesses and support systems throughout a structure.

3.5 Grillage analysis

Grillage analysis is probably the most popular computer aided method for analyzing bridge decks. This is because it is easy to comprehend and use, relatively inexpensive, and has been proved to be reliably accurate for a wide variety of bridge types. The method, pioneered for computer use by Lightfoot and Sawko [6] represents the deck by an equivalent grillage of beams as in Fig. 3.8. The dispersed bending and torsion stiffnesses in every region of the slab are assumed for purpose of analysis to be concentrated in the nearest equivalent grillage beam. The slab's longitudinal stiffnesses are concentrated in the longitudinal beams while the transverse stiffnesses are concentrated in the transverse beams. Ideally the beam stiffnesses should be such that when prototype slab and equivalent grillage are subjected to identical loads, the two structures should deflect identically and the moments, shear forces and torsions in any grillage beam should equal the resultants of the stresses on the cross-section of the part of the slab the beam represents. This ideal can in fact only be approximated to because of the different characteristics of the two types of structure summarized below.

Firstly, equilibrium of any element of the slab requires that torques are identical in orthogonal directions, as Equation 3.4, and also the twist $\partial^2 w/\partial x \partial y$ is the same in orthogonal directions. In the equivalent grillage there is no

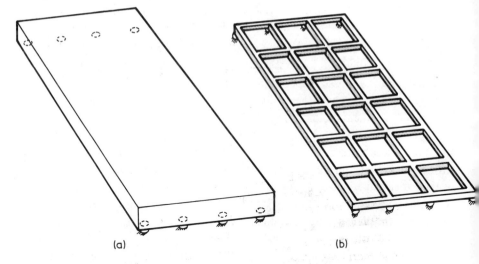

Fig. 3.8 (a) Prototype deck and (b) equivalent grillage.

physical or mathematical principal that makes torques or twists automatically identical in orthogonal directions at a joint. However, if the grillage mesh is sufficiently fine as in Fig. 3.9a, the grillage deflects in a smooth surface with twists in orthogonal directions approximately equal (as will be the torques if torsion stiffnesses are the same in the two directions). On the other hand if the mesh is too coarse as in (b), the grillage will not deflect in a smooth surface so that twists and torques are not necessarily similar in orthogonal directions. Even so it is often found that a coarse mesh is sufficient for design purposes.

Another shortcoming of the grillage is that the moment in any beam is solely proportional to the curvature in it, while in the prototype slab the moment in any direction depends, as in Equation 3.5, on the curvatures in that direction

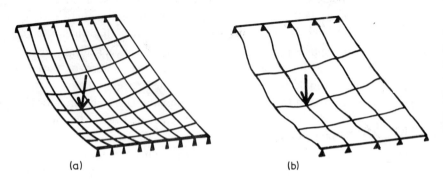

Fig. 3.9 (a) Fine and (b) coarse grillage meshes.

Slab Decks: Grillage Analysis

and the orthogonal direction. Fortunately it has been found from numerous comparisons of grillage with experiments and more rigorous methods that bending stresses deduced from grillage results for distributed moments are sufficiently accurate for most design purposes. In the immediate neighbourhood of a load, which is concentrated in an area much smaller than the grillage mesh, the grillage cannot give the high local moments and torques, and influence charts described in Section 3.8 are necessary.

3.5.1 Grillage mesh

Because of the enormous variety of deck shapes and support arrangements, it is difficult to make precise general rules for choosing a grillage mesh. However it is worth summarizing some of the deck and load characteristics that should be born in mind. Other advice and examples are given in references [7 and 8].

(1) Consider how the designer wants the structure to behave, and place grillage beams coincident with lines of designed strength (i.e. parallel to prestress or component beams, along edge beams, along lines of strength over bearings, etc.)

(2) Consider how forces distribute within prototype. For example if the deck has cross-section as Fig. 3.10, torsion shear flows are as shown. The vertical shear flows at the edges of the slab are represented by components S_x of vertical shear forces in edge grillage members. For the prototype/grillage equivalence to be as precise as possible, each edge grillage member should be close to the resultant of the vertical shear flows at edge of deck. For a solid slab this is about 0.3 of depth from the edge.

(3) The total number of longitudinal members can be anything from one (if slab is narrow enough to behave as beam) to twenty or so (if deck is very wide

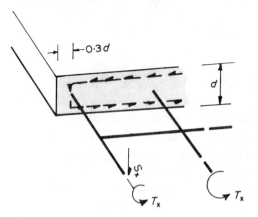

Fig. 3.10   Torsion forces at edge of grillage.

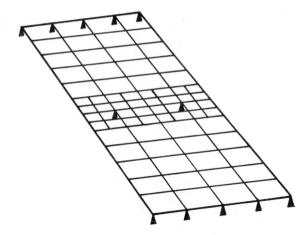

Fig. 3.11 Fine grillage mesh in region of sudden change over internal support.

and design critical enough to warrant expense and trouble). There is little point in placing members closer than 2 to 3 times slab depth since local dispersion of load within slab is not considered. On the other hand, if output information is to illustrate local high values, the maximum separation of longitudinal members for isotropic slabs should not be more than about ¼ effective span. For orthotropic slabs, the charts of Chapter 10 can be used to adopt a spacing so that a member with point load above carries not more than 40 per cent of the load.

(4) The spacing of transverse members should be sufficiently small for loads distributed along longitudinal members to be represented with reasonable accuracy by a number of point loads, i.e. spacing less than about ¼ effective span. In regions of sudden change such as over internal support, a closer spacing is necessary, as in Fig. 3.11.

(5) The transverse and longitudinal member spacings should be reasonably similar to permit sensible statical distribution of loads.

(6) Simply supported decks at skew angles less than 20° can usually be analyzed with grillages having right supports. However for a higher angle of skew, or if the deck is continuous, the lines of the grillage supports should be within about 5° of the skew supports of the prototype.

(7) In general, transverse grillage members should be at right angles to longitudinal members (even for skew bridges) unless, as discussed in Chapter 9, directions of strength such as reinforcements are skew.

(8) If deck is at high skew or bearings are close together, the compressibility of the bearings has considerable effect on local shear forces, etc. and so should be represented with care.

(9) It is implicitly assumed in a grillage analysis that point loads represent loads distributed over the width represented by the member. Sometimes decks with isolated point supports are best studied with two independent grillages. The first, with a coarse mesh of the whole deck, is used to study distribution of moments, etc. between spans: the second, with a finer mesh, represents only a small region around the support. The forces and displacements applied to the boundaries of this smaller grillage are derived from the forces and displacements output for the same points in the coarse grillage.

3.5.2 Grillage member section properties

Bending inertias

The bending inertia of longitudinal and transverse grillage members are calculated by considering each member as representing the deck width to mid-way to adjacent parallel members as shown in Fig. 3.12. The moment of inertia is calculated about the neutral axis of the deck. Thus for an isotropic slab,

$$I = \frac{bd^3}{12}. \qquad (3.14)$$

If the deck has thin cantilever or intermediate slab strips as Fig. 3.13, the longitudinal members can be placed as in (a) or as in (b). In (a) the inertias of all members are calculated about the deck neutral axis. However if the grillage members are placed as in (b), the thin slabs above members 1, 5 and 9 act primarily as flanges to members 2, 4, 6 and 8 respectively. Consequently the inertias of 1, 5 and 9 are calculated about centroid of thin slab, while members 2, 4, 6 and 8 are calculated with the flanges as in (a) but with small inertias of 1, 5 and 9 deducted. Transversely, the thin slab flexes about its own centroid so that the thin slab depth is used in Equation 3.4 for members 1–2, 4–5, 5–6, 8–9 while the thick slab depth is used for 2–3, 3–4, 6–7, and 7–8.

For a voided slab deck such as Fig. 3.14 the longitudinal grillage member inertias are calculated for shaded section about neutral axis. Transversely, the

Fig. 3.12 Subdivision of slab deck cross-section for longitudinal grillage members.

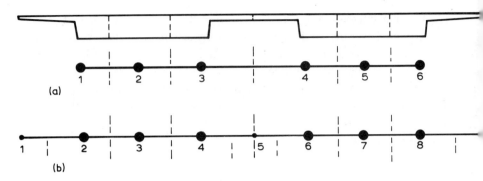

Fig. 3.13 Alternative positions for longitudinal grillage members for deck with thin cantilever and connecting slabs.

inertia is generally calculated as at the centre line of void. However for void depths less than 60 per cent of the overall depth, the transverse inertia can usually be assumed to be equal to the longitudinal inertia per unit width. Neither calculation is precise, but both are sufficient for design purposes.

If the moment curvature Equation 3.5 for a slab is compared with Equation 2.3 for a beam, it can be seen that the slab equation differs not only due to effect of transverse curvature but also because the effective stiffness is $1/(1 - \nu^2)$ times that of the beam. This factor of increased stiffness of slab over equivalent beam is generally ignored in grillage analysis because both longitudinal and transverse stiffnesses are affected by the same relative amount and so do not alter distributions of load.

Reinforced and prestressed concrete slab bridges often have similar stiffnesses in longitudinal and transverse directions with the result that sufficient accuracy is obtained assuming that the full uncracked concrete section is effective, with reinforcing steel ignored. However if the transverse reinforcement is light while longitudinally the bridge is prestressed or heavily reinforced, account should be taken of flexural cracking, and the inertias in the two directions calculated separately for the different transformed sections.

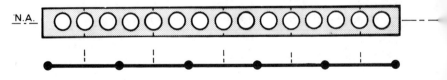

Fig. 3.14 Positions of longitudinal grillage members for voided slab decks.

Torsion

In Section 3.3.2 it was shown that the torsion constant per unit width of a slab is given by

$$c = \frac{d^3}{6} \text{ per unit width.}$$

Thus for a grillage beam representing width b of slab,

$$C = \frac{bd^3}{6}.$$

This is twice the magnitude of the moment of inertia given by Equation 3.14, and in general it is possible to assume $C = 2I$ for grillage members representing slabs. There is no simple rigorous rule for calculating C for voided slabs and the above rule of $C = 2I$ is as convenient and accurate as any.

In true orthotropic slabs, the torques in transverse and longitudinal directions are equal by Equation 3.4 and at the same time both twists are identically equal to $\partial^2 w/\partial x \partial y$. Consequently the transverse and longitudinal grillage members should have identical torsion constants per unit width of deck. Following the approximation of Huber in Equation 3.9 it is suggested that the torsion constant of transverse and longitudinal grillage beams be

$$c = 2\sqrt{(i_x i_y)}$$

where c = torsion constant per unit width of slab, i_x = longitudinal member inertia per unit width of slab and i_y = transverse member inertia per unit width of slab.

In beam-and-slab construction and 'orthotropic' steel battledeck construction, torques are not the same orthogonally and Equation 3.16 does not apply (see Chapter 4 and [2]), and tortion constants are different.

At the edges of a slab, the resultant horizontal shear flows near top and bottom faces (Fig. 3.10) terminate short of the edge of the slab at a distance of approximately 0.3 of depth. The equivalence of grillage and prototype is improved if the width of edge member is reduced for calculations of C to $(b - 0.3d)$.

It was mentioned in Section 3.3.2 that 'torque' in a slab describes only the torque on a section due to opposed horizontal shear flows near top and bottom faces, while the vertical shear flows at the edges are considered as part of vertical shear forces. The grillage reproduces the behaviour very closely. Fig. 3.15 shows the slab of Fig. 3.5 in (a) together with equivalent grillage in (b). The forces on the cross-section in (b) are equivalent to those in (a) with grillage member

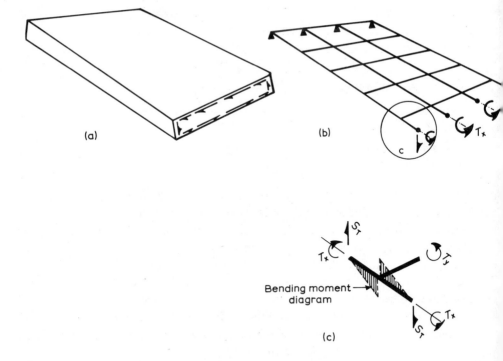

Fig. 3.15 Torsion of (a) slab, (b) equivalent grillage and (c) forces in part of grillage.

torques T_x equivalent to torque in slab due to opposed horizontal shear flows while shear forces S_T are equivalent to vertical shear forces in slab. The reason the shear force S_T is generated in edge members of grillage (and at edges of slab) by torsion is demonstrated in (c). Twisting of the grillage induces twisting and torques in both longitudinal and transverse members. At the joint between a transverse member and the edge longitudinal member, the transverse torque reacts with bending moments and shear forces S_T in the longitudinal member. At internal joints most of the transverse torque passes across the joint, and only the small difference in transverse torques on the two sides reacts with bending and shear longitudinally.

One exception to the above rule occurs if the grillage for a beam-like bridge is devised with only one structural longitudinal member and various outrigged transverse members. Since the longitudinal member must carry the whole of the torque on cross-section due to opposed horizontal shear flows and opposed vertical edge shear forces, the torsion constant must be calculated as for a beam in Section 2.

3.6 Grillage examples

3.6.1 Solid slab

Fig. 3.16 shows a single span solid reinforced concrete slab deck and a convenient grillage mesh. Since the grillage is small, as is its cost, it is worth using quite a fine mesh here with transverse and longitudinal members at spacing equal to approximately 2½ times the depth. Edge longitudinal members are located at distance of 0.3 of depth from edge to be close to position in prototype of vertical shear forces due to torsion.

Since reinforcement in two directions is of the same order of magnitude, the stiffnesses will be assumed equal so that the slab is isotropic. By calculating the inertias on the full concrete area (i.e. ignoring cracking) we obtain

$$i_x = i_y = \frac{1.0^3}{12} = 0.0834 \quad \text{per unit width of slab in member}$$

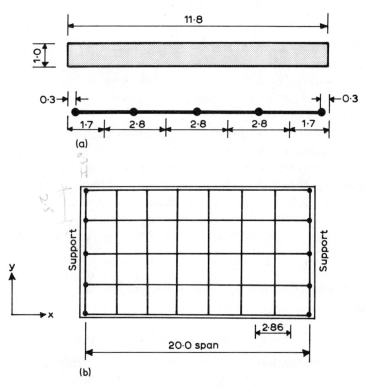

Fig. 3.16 Grillage of solid slab deck. (a) Section (b) plan.

$$c_x = c_y = \frac{1.0^3}{6} = 0.167 \quad \text{per unit width of slab in member.}$$

For internal longitudinal members with widths of 2.8 m we obtain

$$I_x = 2.8 \times 0.0834 = 0.233 \quad C_x = 2.8 \times 0.167 = 0.466.$$

For the edge longitudinal member the width is 1.7 m for calculation of I and $(1.7 - 0.3) = 1.4$ for calculation of C (i.e. width subjected to horizontal torsional shear flows.) Hence

$$I_x = 1.7 \times 0.0834 = 0.142 \quad C_x = 1.4 \times 0.167 = 0.233.$$

Transverse inertias are calculated in same way.

3.6.2 Composite solid slab with edge stiffening

Fig. 3.17 shows a single span slab deck constructed compositely of precast prestressed concrete beams with infill reinforced concrete. The beams span right to the abutment and at the high-skew edges they are supported by an edge beam forming part of the parapet.

The grillage mesh has been chosen so that longitudinal members are parallel to prestressed beams with the transverse members at right angles. Each longitudinal member represents three prestressed beams, while transverse members are placed at ¼ span spacing. The edge member is concentric with the centre line of the edge stiffening upstand.

The infill concrete has lower strength and stiffness than the prestressed concrete so that is has a modular ratio $m = 0.8$ compared to prestressed concrete. Differing transformed cracked section inertias are used in the two directions because the transverse reinforcement is light. Furthermore, this reinforcement has different areas in top and bottom mats so that the transverse transformed section has different inertias for sagging and hogging; the average of the two is used. As a result with $m = 7$ for reinforcement (short term loading),

$i_x = 0.0070$ per unit width of slab in member

$i_y = 0.000\,35$ per unit width of slab in member.

Applying Equation 3.16,

$c_x = c_y = 2\sqrt{(i_x\, i_y)} = 0.0031$ per unit width of slab in member.

The section properties of the edge stiffening beam are calculated for the area shown crosshatched in Fig. 3.17. (It is assumed that the effect of the composite slab is included in internal orthogonal grillage members.) Since the edge

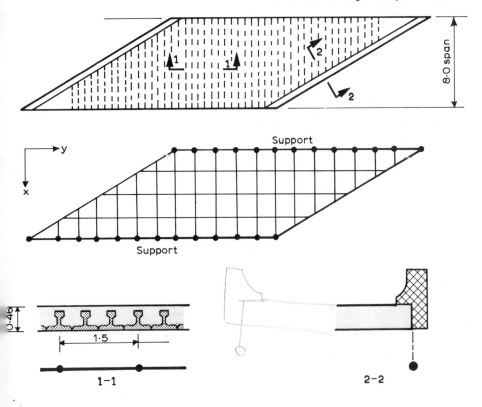

Fig. 3.17 Grillage of skew composite slab deck.

stiffening behaves like a beam, the torsion constant is calculated as per Section 2.4. It should be noted that the inertia of the edge beam has not been increased here in the manner described in Section 8.3 because the transverse stiffness of the composite slab may not be sufficient to act as an effective flange to the upstand.

3.6.3 Two span voided slab

Fig. 3.18 shows a two span voided slab deck with edge cantilever slabs supporting services. The grillage has four longitudinal members. The two-edge members are concentric with centre line of edge webs in which torsion vertical shear flows will be concentrated. Internal longitudinal members pass through bearing positions. The transverse members are in general orthogonal to the longitudinal members and at approximately 1/5 effective span centres. Near the internal support they are closer to permit analysis of the sudden variations. In

66 *Bridge Deck Behaviour*

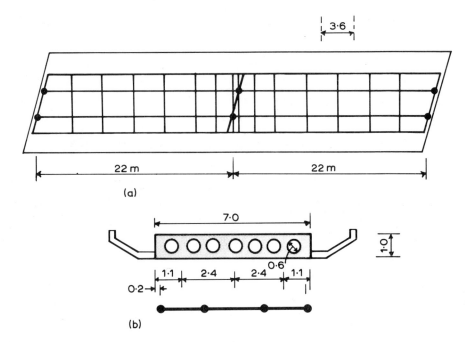

Fig. 3.18 Grillage of continuous voided slab deck. (a) Plan (b) section.

addition a skew member is placed between the bearings to represent the concentration of strength in the form of diaphragm beam reinforcement.

The longitudinal inertia is calculated for the shaded area in Fig. 3.16 which gives

$$i_x = \frac{1.0^3}{12} - \frac{\pi \times 0.6^4}{64} = 0.077 \quad \text{per unit width of slab in member.}$$

Assuming the deck behaves isotropically,

$i_y = i_x$

$c_x = c_y = 2i_x = 0.154$ per unit width of slab in member.

Thus we find for the internal longitudinal members

$I_x = 2.4 \times 0.077 = 0.185 \quad C_x = 2.4 \times 0.154 = 0.37$

while for the edge members

$I_x = 1.1 \times 0.077 = 0.085 \quad C_x = 0.9 \times 0.154 = 0.14.$

It will be seen that the torsion constant is calculated only for the width 'inside' the edge vertical shear forces (here coincident with edge member). For orthogonal transverse edge members near mid-span

$I_y = 3.6 \times 0.077 = 0.277 \quad C_y = 3.6 \times 0.154 = 0.55.$

Determination of the section properties of the skew diaphragms is imprecise. It is suggested that it should be based on the amount of reinforcement in the diaphragm in excess of that distributed as elsewhere in the slab.

3.7 Interpretation of output

3.7.1 Bending moments and shear forces

Fig. 3.19 shows the typical shape of bending moment diagrams for three longitudinal members of a grillage near one edge. The diagram for internal members is usually reasonably continuous and the design moments can be read straight off the grillage output. The edge member diagram is typically discontinuous with 'saw teeth' because of the effects of torsion illustrated in Fig. 3.15. The saw tooth diagram of Fig. 3.19 can be thought of as the superposition of a saw tooth moment diagram due to torsion on a continuous moment diagram, shown dotted, due to bending. As a result, the bending moments should be taken as the average of grillage output moments on each side of a node.

The shear force output for each grillage member is the slope of the output saw tooth bending moment diagram. It includes both the component due to true bending S_M and the component due to torsion S_T as shown in Fig. 3.19. Since

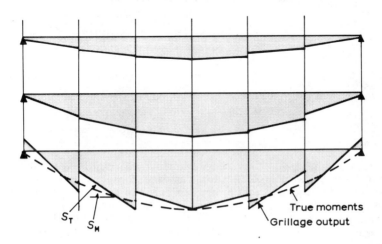

Fig. 3.19 Part of grillage output moment diagram.

both components act together in the prototype slab it should be designed to support the full bending and torsional shear force as output by the grillage.

The torque in a true orthotropic slab is the same in orthogonal directions, however it is often different when read from grillage output. The design torque at any point should be taken as the average of those output for local transverse and longitudinal members per unit width of slab.

3.8 Moments under concentrated loads

The effective area of application of a concentrated load can be assumed to spread at 45° to the vertical down through the surfacing and slab to the plane of the neutral axis as shown in Fig. 3.20. If this area of application is equal to or larger than the grillage mesh (or if the areas due to several loads touch and together are larger), the load can be assumed to be sufficiently dispersed for the grillage to reproduce the distribution of moments throughout the slab. No further modification of moments, etc. is necessary. On the other hand, if the area of application of the load is small compared to the grillage mesh, no information will be obtained about the local high values under the load, though the grillage distributed moment field will simulate that in the deck. The additional moments due to high local curvatures can be obtained for the area of slab within grillage mesh from influence charts such as those of Pucher [5].

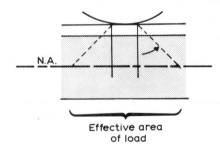

Fig. 3.20 Dispersion of concentrated load to plane of neutral axis.

REFERENCES

1. Troitsky, M. S. (1967), *Orthotropic Bridges: Theory and Design*, The James F. Lincoln Arc Welding Foundation, Cleveland, Ohio.
2. American Institute of Steel Construction (1963), *Design Manual for Orthotropic Steel Plate Deck Bridges*, AISC, New York.
3. Rowe, R. E. (1962), *Concrete Bridge Design*, C. R. Books, London.
4. Jaeger, L. G. (1964), *Elementary Theory of Elastic Plates*, Pergamon Press, Oxford.

5. Pucher, A. (1964), *Influence Surfaces of Elastic Plates*, Springer Verlag, Wien and New York.
6. Lightfoot, E. and Sawko, F. (1959), 'Structural frame analysis by electronic computer: grid frameworks resolved by generalised slope deflection,' *Engineering*, **187**, 18–20.
7. West, R. (1973), *C & CA/CIRIA Recommendations on the Use of Grillage Analysis for Slab and Pseudo-slab Bridge Decks*, Cement and Concrete Association, London.
8. West, R. (1973), 'The use of grillage analogy for the analysis of slab and pseudo-slab bridge decks,' Research report 21, Cement and Concrete Association, London.

4
Beam-and-slab decks : grillage analysis

4.1 Introduction

A large proportion of modern small and medium span bridges have beam-and-slab decks. This chapter describes how for the purposes of design they can be thought of as two-dimensional structures with behaviour which is in several ways simpler than that of slabs. They lend themselves to analysis by computer aided grillage with an equivalence of prototype and model which has immediate appeal to engineers. Examples are demonstrated for several types of structure. Occasionally for bridges which are not within the range of common practice it is necessary to investigate the secondary characteristics of such decks resulting from their truly three-dimensional form. These are mentioned at the end of the chapter and investigated further in Chapter 7.

4.2 Types of structure

The majority of beam-and-slab decks have a number of beams spanning longitudinally between abutments with a thin slab spanning transversely across the top as shown in Fig. 4.1. For short spans the beams are often contiguous as in (a), but for longer spans the beams are spaced as in (b) and (c). Transverse beams, called 'diaphragms', are placed to connect the longitudinal beams over the supports and sometimes at various sections along the span as in (d). The deck

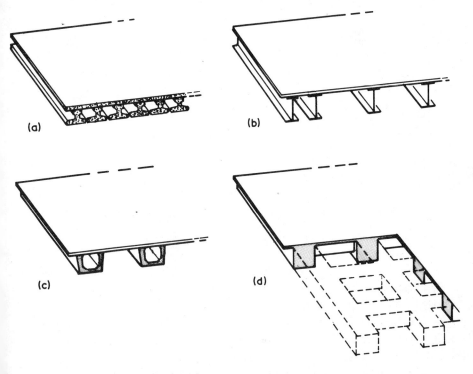

Fig. 4.1 Beam-and-slab decks. (a) Contiguous (b) spaced I-beams
(c) spaced box beams (d) grid.

can have a high skew angle with beams displaced longitudinally relative to each other, and can be tapered with the beams not parallel. Curvature of the supported carriageway is usually accommodated by giving the edges of the slab the appropriate curvatures while supporting it on beams which are straight for each span. However, sometimes beams are also curved as described in Chapter 9.

4.3 Structural action

The behaviour of a beam-and-slab deck without mid-span diaphragms as shown in Fig. 4.2 can be thought of as a simple combination of beams spanning longitudinally and slab spanning transversely. For longitudinal bending, the slab acts as top flange of the beams, and the deck can be thought of (and is sometimes constructed) as a number of T-beams connected along the edges of the flanges. Since the slab has a bending stiffness only a fraction of that of the beams, it flexes with much greater curvatures transversely than longitudinally, and in spanning between beams behaves much like a large number of transverse

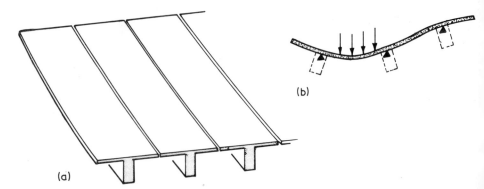

Fig. 4.2   Action of slab of beam-and-slab deck in (a) longitudinal bending as flanges of T-beams and (b) transverse bending as continuous beam.

spanning planks. It is only in the immediate neighbourhood of a concentrated load that longitudinal moments and torques in the slab are of comparable magnitude to transverse moments. As shown in Section 4.6, it is generally possible to superpose the moments due to local two-dimensional dispersion of concentrated load on the transverse moments in the slab related to relative deflections and rotations of the beams.

Fig. 4.3 shows an element of deck supporting an element dW of the locally dispersed load. The beam transmits moment M_x shear force S_x and torsion T_x while the slab effectively only transmits transverse moment m_y and shear force

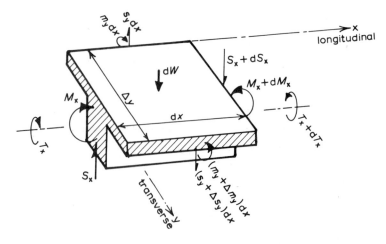

Fig. 4.3   Element of beam-and-slab deck.

s_y (per unit width of slab). The forces are related by equations

$$\frac{dS_x}{dx} + \Delta s_y = -W\Delta y$$

$$\frac{dM_x}{dx} = S_x \tag{4.1}$$

$$\Delta m_y + \frac{dT_x}{dx} = s_y \Delta y.$$

The torsion in the slab has been omitted since it is comparatively small in such a thin slab. If, in addition, the beams have very thin I-sections their torsional stiffness is also very low so that T_x is effectively zero. The slab is then similar to a continuous transverse beam supported on an elastic support at each beam. In contrast, if the beams have high torsional stiffness, T_x is significant and the moments in the slab are discontinuous over the beam.

When the deck has transverse beams as in Fig. 4.4 the forces are related by

$$\Delta S_x + \Delta S_y = -W\Delta x\Delta y$$
$$\Delta M_x + \Delta T_y = S_x \Delta x \tag{4.2}$$
$$\Delta M_y + \Delta T_x = S_y \Delta y.$$

These equations resemble those for a slab in Section 3.2. However the torques

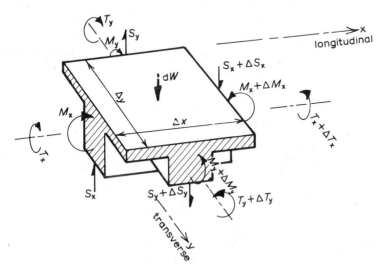

Fig. 4.4 Element of grid deck or of beam-and-slab deck at diaphragm.

74 *Bridge Deck Behaviour*

T in the two directions are not equal here, and depend on the differing twists and stiffnesses in the two directions.

In general, the stiffnesses of parts of the structure in longitudinal and transverse directions can be assumed the same as beams of similar cross-section, and stresses calculated from actions using Sections 2.3.2 and 2.4.3. Special considerations relating to slab behaviour and effective width of slab acting as flange to beams are discussed in next section.

4.4 Grillage analysis

4.4.1 Grillage mesh

Determination of a suitable grillage mesh for a beam-and-slab deck is, as for a slab deck, best approached from a consideration of the structural behaviour of the particular deck rather than from the application of a set of rules. Fig. 4.5 shows four examples of suitable meshes for four types of deck; Section 4.5 and references [1 and 2] give further examples.

In Fig. 4.5a the deck is virtually a grid of longitudinal and transverse beams. Since the average longitudinal and transverse bending stiffnesses are comparable, the distribution of load is somewhat similar to that of a torsionally flexible slab, but with forces locally concentrated. The grillage simulates the prototype closely by having its members coincident with the centre lines of the prototypes beams.

The deck in Fig. 4.5b has longitudinal beams at centres a little less than the lane widths, and it is both convenient and physically reasonable to place longitudinal grillage members coincident with the centre lines of the prototype's beams. With no midspan transverse diaphragms the spacing of transverse grillage members is somewhat arbitrary, but about $1/4$ to $1/8$ of effective span is generally convenient. Where there is a diaphragm in the prototype such as over a support, then a grillage member should be coincident.

Fig. 4.5c is a deck with contiguous beams at very close centres. Since a grillage with longitudinal members coincident with all beams would be very expensive and unmanageable, it is expedient to represent more than one beam by each longitudinal grillage member. However as beam-and-slab decks have poor distribution characteristics, it is important not to place longitudinal grillage members much further apart than about 1/10 span, otherwise the concentration of moment will not be apparent in the grillage analysis. See also spacing recommendations of Section 10.5.2.

The deck of Fig. 4.5d has large beams whose widths form a significant fraction of the distance between the centre lines. Since during transverse flexure the thin slab flexes much more than the thick beams, the grillage must also flex

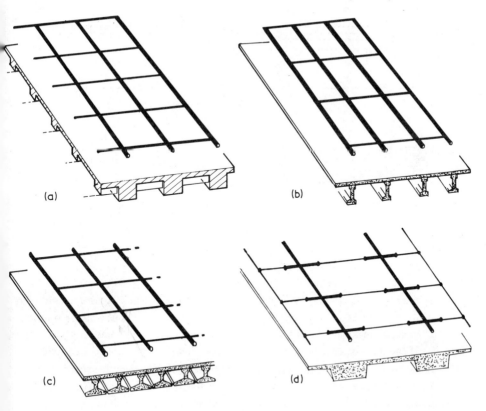

Fig. 4.5 Grillage meshes.

with most of the bending over the width of the thin slab. Accordingly the transverse members are made up of a string of members whose different stiffnesses represent the different stiffnesses in the prototype. (An alternative approach is to use shear flexibility as described in Section 6.3 so that a single transverse member between longitudinal beam centre lines is able to represent the thick-thin-thick slab of the prototype. However, processing of the grillage output is then much more cumbersome.) This deck could also be treated as a slab deck as in Fig. 3.13 with two longitudinal members per beam, but then torsion parameters must be calculated as for a slab.

4.4.2 Longitudinal grillage member section properties

Fig. 4.6 shows part cross-sections of three beam-and-slab decks and the amount of each deck represented by the appropriate grillage member.

The flexural inertia of each grillage member is calculated about the centroid

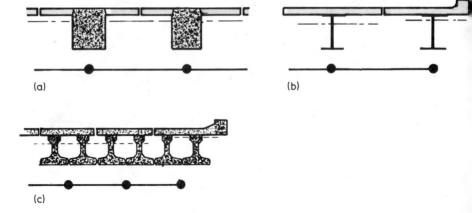

Fig. 4.6 Sections represented by longitudinal grillage members.

of the section it represents. Often the levels of the centroids of internal and edge member sections are at different levels. The significance of this is ignored unless a three-dimensional analysis is performed as described in Chapter 7.

If the deck beams are spaced further apart than 1/6 of the effective span, or if the edge cantilever exceeds 1/12 of the effective span, shear lag significantly reduces the effective width of slab acting as flange to each beam. The grillage inertia should then be calculated using a reduced width of slab as described in Chapter 8.

Sometimes for the purpose of improving the simulation of applied loading demonstrated in Section 4.6, it is convenient to place longitudinal grillage members of nominal stiffness between those representing the structural sections of Fig. 4.6. The section properties of these members are calculated in a similar manner to that outlined in Section 3.5.2 for the deck in Fig. 3.13.

When the various decks in Fig. 4.6 are subjected to torsion, the 'beam' parts (dark shading in Fig. 4.6) behave like beams subjected solely to longitudinal torsion, while the 'slab' parts behave like slabs with torsion in both directions. Consequently the torsion constant C of the grillage member is the sum of the torsion constant of the beam calculated as per Section 2.4.3 and the torsion constant of the slab calculated as per Equation 3.15.

4.4.3 Transverse grillage member section properties

The section properties of a transverse grillage member, which solely represents slab, are calculated as for a slab. For this

$$I = \frac{bd^3}{12}$$

$$C = \frac{bd^3}{6} \tag{4.3}$$

When the grillage member also includes a diaphragm, an estimate must be made of the width of slab acting as flange. If the diaphragms are at close centres, the flanges of each can be assumed to extend to midway between diaphragms. However if these flanges are wider than 1/12 of the effective transverse span between points of zero transverse moment, shear lag reduces the effective flange width as described in Chapter 8. Without prior knowledge of transverse moments, it is usually conservative to assume that the effective flange is 0.3 of distance between longitudinal members (i.e. the span in shear lag calculation is twice the distance between longitudinal members).

If the construction materials have different properties in the longitudinal and transverse directions, care must be taken to estimate their relative stiffnesses. For example, a reinforced concrete slab on prestressed concrete or steel beams could be fully effective in compression for longitudinal sagging moments but behave as transformed cracked section for transverse bending. Furthermore it is possible that if the deck is continuous over a support, the same slab could be cracked for full depth in tension for longitudinal bending over the support so that only the reinforcing steel is effective. While an effort should be made to represent these differing characteristics, precise estimates of stiffness are seldom possible because of the unpredictable inelastic behaviour of construction materials. Information on torsion stiffness of cacked reinforced concrete is given in reference [3].

4.5 Grillage examples

4.5.1 Contiguous beam-and-slab deck

Fig. 4.7 gives details of a single span skew deck constructed of contiguous prestressed precast concrete beams with *in situ* reinforced concrete slab. The grillage model has one longitudinal beam to each two prototype beams. While sufficient transverse members are provided for detailed analysis, their precise positions are chosen so that they intersect support beams at the same points as longitudinal beams. The skew of the grillage, but not the span, differs slightly from that of prototype to make the mesh regular. (Such expediency greatly reduces the risk of human errors in calculation and thus is likely to improve accuracy of analysis.)

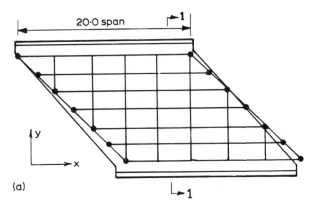
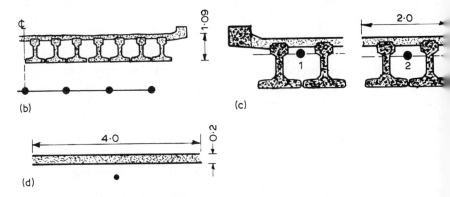

Fig. 4.7 Grillage of skew contiguous beam-and-slab deck. (a) Plan (b) part section 1-1 (c) longitudinal members (d) transverse member.

The longitudinal members are calculated for sections in Fig. 4.7c assuming full concrete areas effective, but with *in situ* slab transformed by modular ratio $m = 0.85$ of Young's Moduli for *in situ* and prestressed concrete.

$$I_{x1} = 0.24 \quad I_{x2} = 0.174.$$

The torsion constants are calculated separately for 'beam' and 'slab' parts of each member and added to give

$$C_{x1} = 2 \times 0.004 + \frac{2.0 \times 0.2^2 \times 0.85}{6} + 0.006 = 0.016$$

$$C_{x2} = 2 \times 0.004 + \frac{2.0 \times 0.2^3}{6} \times 0.85 = 0.010.$$

Beam-and-Slab Decks: Grillage Analysis

The percentage area of reinforcement spanning transversely in such a slab is usually quite high and the inertia calculated on the transformed cracked section does not differ very much from inertia calculated on uncracked section ignoring reinforcement. Consequently, since initially area of reinforcement is not known, the inertia is calculated on uncracked section with $m = 0.85$. Transverse grillage members then have

$$I_y = \frac{4.0 \times 0.2^3}{12} \times 0.85 = 0.0023 \quad C_y = \frac{4.0 \times 0.2^3}{6} \times 0.85 = 0.0045.$$

The torsion constant of the cracked concrete is likely to be in error by an unknown amount, but here has little effect.

The percentage area of reinforcement in the support diaphragm is low, and so the inertia is calculated on the transformed cracked section, with the effect of slab as flange ignored. The torsion constant of such a beam is also very low, as it is not prestressed and it is made up of discontinuous sections of *in situ* concrete and precast beam web. Without relevant experimental evidence, it is suggested that C is calculated for the area of uncracked concrete used in calculating I.

4.5.2 Spaced steel I-beams with reinforced concrete slab

Fig. 4.8 shows part of a composite deck constructed of reinforced concrete slab on steel beams. Longitudinal grillage members are placed coincident with centre lines of steel beams, and each represents the part of deck section shown in (b).

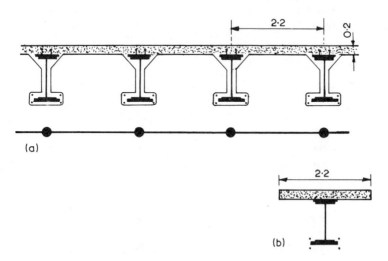

Fig. 4.8 (a) Cross-section of composite steel/concrete deck and of grillage (b) section represented by longitudinal grillage member.

Using modular ratio $m = 7$ for steel (short term loading) and ignoring concrete cracked in tension that encases steel bottom flange, we obtain

$$I_x = 0.21$$

$$C_x = 0.000\,031 \times 7 + \frac{2.2 \times 0.2^3}{6} = 0.0032.$$

The slab is similar to that in Fig. 4.7 so that transverse grillage member properties are calculated in the same way.

4.5.3 Spaced box beam with slab deck

Fig. 4.9 shows the cross-section of a beam-and-slab deck constructed of spaced prestressed precast concrete box beams supporting a reinforced concrete slab. Longitudinal grillage members are placed coincident with centre lines of beams, with additional 'nominal' members running along centre lines of slab strips.

The section properties of the nominal members are calculated for width of slab to midway to neighbouring beams hence

$$I_x = 1.4 \times \frac{0.25^3}{12} = 0.0018 \quad C_x = \frac{1.4 \times 0.25^3}{6} = 0.0036.$$

The properties of the beam members are calculated for the sections with flanges including the area in nominal members (unless shear lag has reduced the effective width of flanges to less), but with previously calculated properties of 'nominal' members deducted:

$$I_x = 0.57 - 2 \times \frac{0.0018}{2} = 0.57$$

$$C_x = 0.34 - 2 \times \frac{0.0036}{2} = 0.34.$$

Transverse members are calculated as in previous examples.

If the beams are much wider than those in Fig. 4.9 in comparison with the beam spacing, account must be taken of the variation in transverse flexural stiffness between slab and beam. If the beams have thick walls so that cells do not distort, a grillage similar to Fig. 4.5d can be used. However if the walls are thin, distortion of the cross-section must be considered. For decks with a few large cells it is simplest to use the techniques described in Chapter 5 for cellular decks using one longitudinal grillage member for each web and shear flexibility to simulate cell distortion. On the other hand, if the deck has a large number of

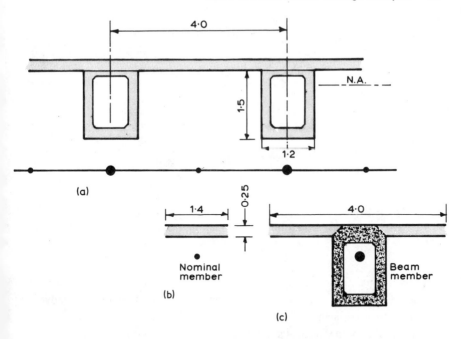

Fig. 4.9 (a) Cross-section of deck with spaced beams and of grillage with nominal members (b) nominal member (c) beam member.

beams it may be simplest to treat it as a shear key deck as in Section 6.3, but using a plane frame analysis (see Fig. 5.15) to determine equivalent transverse grillage properties.

4.6 Application of load

A load standing between beams, as in Fig. 4.10a, can only be represented in the grillage equations in computer program by forces and couples at the joints. Some computer programs carry out a simple local statical distribution of load as is shown in (b). Unfortunately beam-and-slab decks can be very sensitive to such transverse movement of load, and the overall deck deformation and distribution of moments can be markedly different for load cases of (a) and (b). Ideally, the load should be applied to the transverse members or to the joints as transverse fixed edge shear forces and moments given in Equations 2.9. However, some computer load generation routines only do statical distribution, and manual calculations of fixed edge forces is laborious. In such cases it is best to ensure that the transverse spacing of grillage members is less than about 3/4 of the abnormal vehicle or lane load width. Often this can be done by placing 'nominal'

82 Bridge Deck Behaviour

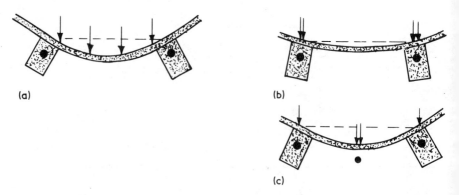

Fig. 4.10 Errors due to statical distributions of loading. (a) Loading (b) erroneous statical redistribution (c) improved statical redistribution with nominal member.

longitudinal members midway between beam members as shown in (c) (see also Section 4.5.3). Statical distribution of loads transversely does not then reduce twisting forces on beams.

The local two-dimensional system of moments and torques in the thin slab under a concentrated load are not given by the grillage and must be obtained either from equations of Westergaard (quoted in Rowe [4]) or more simply from influence charts of Pucher [5]. To apply these methods, the slab strip between beams is assumed to have a span equal to the clear distance between edges of beams plus the effective depth of the slab. Furthermore, the slab is assumed to be fixed along its edges. The effective area of the concentrated load can be assumed to spread at 45° to the vertical down through the surfacing and slab to the mid plane of the slab. The local moments obtained from the charts must be added to those in the slab resulting from twisting and relative deflection of supporting beams. If there are no 'nominal' grillage members between beams and if transverse members have not been loaded, these moments can be read directly from grillage output for the local transverse member. If there is a 'nominal' longitudinal grillage member under the load or if transverse members have been loaded, the slab moments due to twisting of beams are best calculated from grillage output displacements and rotations of adjacent beams using Equations 2.8.

4.7 Interpretation of output

Fig. 4.11 shows a typical bending moment diagram for part of a longitudinal beam of a beam-and-slab deck. Where transverse members represent only thin

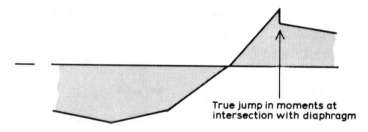

Fig. 4.11 Jump in grillage moment diagram for beam-and-slab deck at intersection of longitudinal beam and transverse diaphragm.

slab, the discontinuities in moment due to transverse torques in the slab are small. The design moment should there be taken as the average of the moments on two sides of joint (as for slab deck in Section 3.7). In contrast, where the transverse member represents a diaphragm beam with significant torsion stiffness, the discontinuity in longitudinal moments is larger and represents a real change in moment across the joint. The design moments should then be taken as different on two sides of the joint and equal to grillage output moments. In a similar way, discontinuities in the moment diagram for transverse slab or diaphragm beams represent real changes in moment at connections with torsionally stiff longitudinal beams.

Design shear forces and torsions can be read direct from grillage output without modification.

Where the grillage member stiffness is calculated from properties of two distinct pieces of structure such as 'beam' and 'slab' in Fig 4.6, the output torque (moment or shear) is attributed to each in proportion to its contribution to the particular stiffness. In general, stresses for prototype design should be calculated from the grillage output by applying the equations of Chapter 2 to the section assumed in the derivation of the grillage member properties.

4.8 Slab membrane action in beam-and-slab decks

In the preceeding discussion it has been tacitly assumed that for consideration of longitudinal bending, the slab can be thought of as a series of strips, each forming a top flange of a T-beam. No check has been made that after notionally cutting up the deck the displacements of the parts are compatible, i.e. that the parts can in fact be joined together without additional forces and distortion hitherto not considered.

Fig. 4.12a shows the midspan section of a beam-and-slab deck with exaggerated deflections due to non-uniform loading. (b) shows the composite

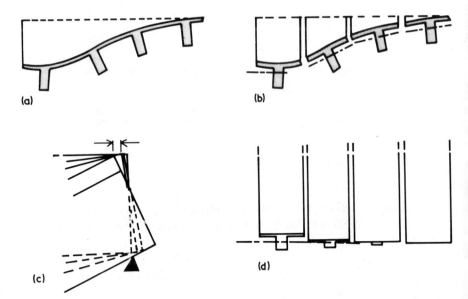

Fig. 4.12 Longitudinal 'warping' movement of slab of beam-and-slab deck. (a and b) span section (c) support elevation (d) support plan.

beams notionally separated, but with twists and deflections of (a). The grillage can adequately simulate these deflections and accompanying transfer of load by vertical shear and transverse bending of the slab. But inspection of the ends of the separate beams in plan or elevation, as in (c) and (d) shows that if all the beams flex about a neutral axis passing through their centroids, the ends of the slab flanges are displaced relative to each other. In reality this step displacement cannot happen, and the relative movement of the tops of the beams is resisted and reduced by longitudinal shear forces in the connecing slab as shown in Fig. 4.13a. These shear forces are in equilibrium with axial tension/compression forces in beams near midspan shown in (b).

This transfer of shear force between beams with balancing axial forces cannot be simulated in a conventional grillage analysis. The forces have three effects on deck behaviour.

(1) The shear forces in the slab can be much larger than predicted from grillage analysis.

(2) The axial tension forces in the beams with largest deflections (i.e. under the load) cause the neutral axis to rise locally while compression forces elsewhere cause the neutral axis to move down, as shown in Fig. 4.13c.

(3) The load distribution characteristics of the deck are improved. The longitudinal inter-beam shear forces and axial forces in Fig. 4.13 are at different

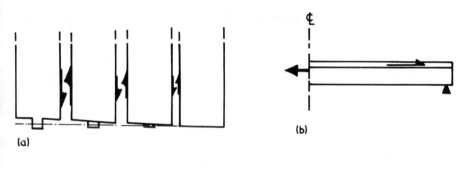

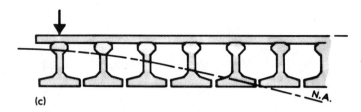

Fig. 4.13 Effects of slab membrane action in beam-and-slab deck. (a) In-plane shear in slab (b) axial force in beam (c) movement of neutral axis.

levels and thus form couples which reduce the moment in the loaded beams and increase moments elsewhere.

It is often assumed that if a deck slab is subjected to shear forces in excess of its strength, designed from grillage analysis, it will only crack or yield and so relax the forces of Fig. 4.13. This may happen, but if the deck is outside the range of common practice, a three-dimensional analysis may be necessary as described in Chapters 7, 12 or 13.

The movement of the neutral axis in Fig. 4.13 occurs as a result of the difference in levels of the centroids of the beams and connecting slabs. The behaviour is further complicated if there are transverse beams also with centroids out of the plane of the slab. However in contrast, the behaviour of cellular decks described in the next chapter is simpler because the centroid does not vary significantly in level between webs and slabs.

REFERENCES

1. West, R. (1973), *C & CA/CIRIA Recommendations on the Use of Grillage Analysis for Slab and Pseudo-slab Bridge Decks*, Cement and Concrete Association, London.
2. West, R. (1973), 'The use of grillage analogy for the analysis of slab and pseudo-slab bridge decks,' Research report 21, Cement and Concrete Association, London.

3. Lampert, P. (1973), Postcracking stiffness of reinforced concrete beams in torsion and bending. *Analysis of Structural Systems for Torsion*, SP-35 American Concrete Institute, Detroit, pp. 384–432.
4. Rowe, R. E. (1962), *Concrete Bridge Design*, C. R. Books, London.
5. Pucher, A. (1964), *Influence Surfaces of Elastic Plates*, Springer Verlag, Wien and New York.

5
Cellular decks : shear-flexible grillage analysis

5.1 Introduction

This chapter describes the modes of deformation and systems of internal forces that characterize the behaviour of cellular decks. It is shown how this behaviour can be investigated using a shear-flexible grillage analysis. The application of this computer aided method pioneered by Sawko [1] is demonstrated for a variety of structural forms. While the shear-flexible grillage is not the most theoretically rigorous analogy of cellular behaviour available, it has the benefit of being applicable to a wide variety of structures, of being relatively inexpensive in computer time and user time, and being relatively simple to comprehend. The shortcomings of the analysis are demonstrated to highlight particular characteristics of cellular structures, and at the end of the chapter other methods are mentioned which might be more appropriate to particular structural details.

5.2 Types of structure

Fig. 5.1 shows a number of cellular structures for which the shear-flexible grillage can be used. The analysis is most appropriate for wide multicellular decks with thin slabs enclosing rectangular cells as shown in Fig. 5.1a. However, it can also be used for decks with only one or a few cells, and has been found to perform with acceptable accuracy for the analysis of decks with inclined webs as

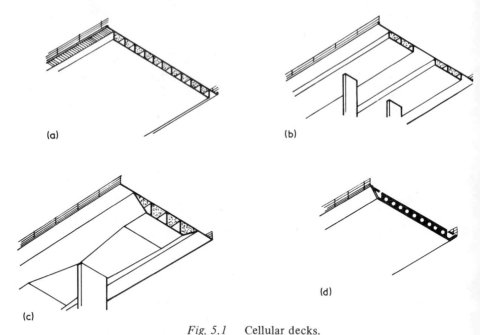

Fig. 5.1 Cellular decks.

in Fig. 5.1c. It can also be used for decks with large cylindrical voids such as Fig. 5.1d. Variations in structural depth or plate thicknesses can be considered, but not the effects of arching action in deep haunches. The decks can also be curved or tapered in plan, but the analysis has the limitations described in Chapter 9. Transverse diaphragms can be placed anywhere at right angles or skew to the longitudinal webs.

5.3 Grillage mesh

The following discussion of cellular behaviour and grillage simulation assumes that the grillage mesh is in the plane of the principal axis of bending of the deck as a whole with the longitudinal members coincident with longitudinal webs. Such positioning of longitudinal members is demonstrated in Fig. 5.2 for the decks of Fig. 5.1. This arrangement is chosen so that the web shear forces can be directly represented by grillage shear forces at the same points on the cross-section. Other arrangements are possible, but then the method of apportioning forces and stiffnesses to members is different from that described later. If the deck has sloping webs, the grillage simulation is not so precise and engineering judgement must be used to position longitudinal members. The members located along the edges of the side cantilevers are not generally necessary for the analysis

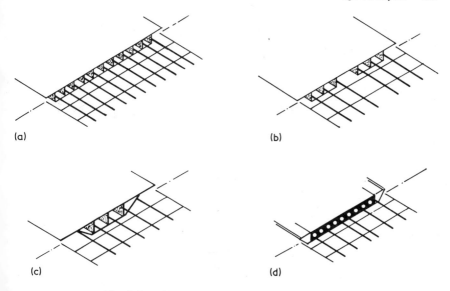

Fig. 5.2 Grillage meshes for cellular decks.

of the cellular structure, but if they are included with nominal stiffnesses they can simplify description of cantilever loading in the computer input. A single cell box can be analysed with a grillage having one longitudinal member to each web. (However if the resultant load is highly eccentric, a three-dimensional space frame (as in Fig. 7.15), folded plate, or finite element analysis may be more appropriate.)

The transverse members, shown in Fig. 5.2, are spaced closer than ¼ of the distance between points of contraflexure and wherever there are diaphragms unless they are too numerous. Wider spacing results in inaccuracy due to excessively large discontinuities in moments, etc. at the joints. Closer spacing results in more continuous structural behaviour and provides greater detail of forces etc., but it does not make the characteristic behaviour of the grillage any closer to that of the cellular structure.

The underlying principle on which grillage member properties are derived is that when the grillage joints are subjected to the same deflections and rotations as coincident points in the structure, the member stiffnesses generate forces which are locally statically equivalent to the forces in the structure.

5.4 Modes of structural action

Fig. 5.3 shows the displacement and deformation of a cross-section of a cellular deck under load split up into the four principle modes: longitudinal bending,

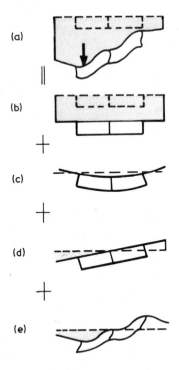

Fig. 5.3 Principal modes of deformation. (a) Total   (b) longitudinal bending   (c) transverse bending   (d) torsion   (e) distortion.

transverse bending, torsion and distortion. The character and simulation of each of these modes is described below.

5.4.1 Longitudinal bending

Longitudinal bending behaviour can be visualized by notionally cutting the deck longitudinally between webs into a number of 'I-beams' as in Fig. 5.4. The longitudinal bending stresses on the cross-section in Fig. 5.5a are similar to those for the 'I-beam' subjected to the same curvature as the deck, and are given by

$$\frac{\sigma}{z} = \frac{M}{I} = \frac{E}{R}. \tag{5.1}$$

The shear stress distribution due to bending is also similar to that from simple beam theory of 'I-beams' with transverse or longitudinal shear flow at the point

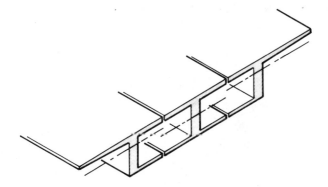

Fig. 5.4   Cellular deck split up into I-beams by cuts along centres of cells.

given by

$$r = \frac{S_M A \bar{z}}{I} \qquad (5.2)$$

where A and $\bar{z}$ are the area and eccentricity of the centre of gravity of the part of the flange beyond the point of consideration. S_M is the vertical shear force in the 'I-beam' (i.e. in the web) due to bending and is given by

$$S_M = \frac{dM}{dx}. \qquad (5.3)$$

S_M is only part of the total vertical shear force in the web which has another component S_T due to torsion.

In Fig. 5.4 the deck is shown cut midway between webs. If the centres of gravity of the beams are at different levels as is the case when the edge beam has a wide top slab cantilever, then on being flexed separately, each will have zero

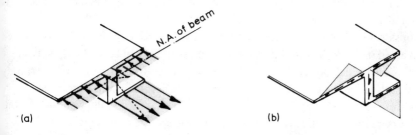

Fig. 5.5   Flexure of edge beam about its own neutral axis. (a) Bending stresses (b) shear flow.

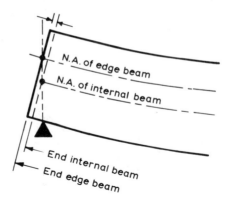

Fig. 5.6 Relative end displacement of beams with different neutral axes.

extension along its own neutral axis at the level of its centre of gravity. Consequently, as shown in Fig. 5.6, cross-sections near the end would rotate about points at different levels and have a relative longitudinal displacement u. This displacement is resisted and reduced by the very high shear stiffnesses of the top and bottom slabs, and all the beams are forced to flex about a neutral axis which is virtually coincident with the principal axis of the deck as a whole. Consequently the section properties of each I-beam represented by a grillage member should be calculated about the principal axis of the deck as a whole. The bending stresses are then calculated from these properties, using Equation 5.1, with z measured from the common neutral axis.

Although the subdivision of the cellular deck into 'I-beams' is somewhat arbitrary, the stresses calculated as above from Equation 5.1 can be considered accurate for practical purposes. A small change of the arbitrary position of the cuts changes the relative magnitudes of the grillage member inertias which in turn attract a different distribution of moments. But the ratio of M/I for any grillage beam remains effectively constant and the calculated bending stresses are not affected. However, shear flows calculated from Equation 5.2 can only be considered approximate since the flange area A is arbitrary. These errors can be avoided if the deck is cut into 'I-beams' whose individual centres of gravity are on the principal axis of the deck. For the example of Fig. 5.7 this is done simply by cutting the slabs so that each web has identical proportions of top and bottom slab. On being flexed separately, the neutral axis of all beams will be on the same level and the distributions of bending stress and shear flow will be as Fig. 5.8. As before, the bending stresses are described by Equation 5.1 where I is the moment of inertia of the asymmetrical-beam section of Fig. 5.7 and z is the distance from the neutral axis of the beam (and deck). Shear flows due to

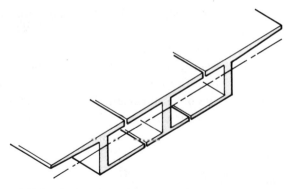

Fig. 5.7 Cellular deck split up into I-beams with common neutral axis.

bending are now accurately described by Equation 5.2 where A, $\bar{z}$ and I relate to the section of Fig. 5.7.

In the above discussion it has been assumed that the I-beams are behaving as shear rigid in response to longitudinal bending. This assumption is reasonable if the flanges are narrow (i.e. if the cell widths and cantilever are narrow). But if the flanges are wide, shear deformation reduces the efficiency of the sections which have a reduced effective width of flange, as is generally accepted in design of T-beams and L-beams. This reduction of efficiency due to shear lag is discussed in Chapter 8. It has also been assumed that the bending compression/tension stresses are uniform across the width of the I-beams. If the beams across the deck are subjected to different curvatures, the bending stresses vary across the deck. This variation is continuous and, ignoring shear lag, can be assumed to be linear between webs. The stresses of Equation 5.1 then represent the average stress across the 'I-beam'.

From the above discussion it is evident that the moments of inertia apportioned to longitudinal grillage members should be calculated about the

Fig. 5.8 Flexure of edge beam with centroid on neutral axis of deck.
(a) Bending stresses   (b) shear flow.

principal axis of the deck as a whole. If possible, the grillage member should be made to represent part of the deck as in Fig. 5.7 which has its centre of gravity on this axis. However for a deck such as Fig. 5.1c, the top slab is very much wider than the bottom slab and such division of the cross-section gives beams of ridiculous cross-sections. In this case it is convenient to notionally cut the deck midway between the positions of longitudinal grillage members and accept some loss of accuracy. The moments of inertia and section moduli of members are still calculated about the principal axis of the deck as a whole.

5.4.2 Transverse bending

Transverse bending, shown in Fig. 5.9, is the flexure of top and bottom slabs in unison about a neutral axis at the level of their common centre of gravity, as if they were connected by a shear rigid web. It does not include the independent flexure of top and bottom slabs which results in cell distortion as shown in Fig. 5.3e.

The moment of inertia of transverse grillage members is calculated about the common centre of gravity of the slabs and is

$$i_t = (h'^2 d' + h''^2 d'') = \frac{h^2 d' d''}{(d' + d'')} \text{ per unit length} \qquad (5.4)$$

where d', d'', h', h'' are the slab thicknesses and distances from their centroid shown in Fig. 5.9. If the transverse grillage member also includes a diaphragm, the inertia should be calculated including the diaphragm.

Grillage analysis ignores the effects of Poisson's ratio on the interaction of longitudinal and transverse moments. In narrow decks this introduces little error as transverse moments are small and the deck is free to hog transversely, just like a beam under a longitudinal sagging moment. However in a wide deck with little stiffness against cell distortion, the calculated transverse moments can be considerably in error if they are small and if Poisson's ratio is significant. The transverse moment due to interaction of the longitudinal moment on the

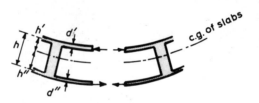

Fig. 5.9   Transverse bending.

transverse moment can be large in comparison to the transverse moment due to transverse sagging. Since concrete has a relatively low Poisson's ratio (of approximately 0.15), its effect is usually ignored.

5.4.3 Torsion

The term 'torsion' as applied to cellular decks describes the shear forces and deformation induced by twisting the deck as in Fig. 5.3d without the effects of distortion of the cross-section in Fig. 5.3e.

When a cellular deck is twisted bodily, there is a network of shear flows round the slabs and down and up the webs as shown in Fig. 5.10a. Most of the shear flow passes round the perimeter slabs and webs, but some short-cuts through intermediate webs. When a grillage is twisted in a similar fashion, the forces crossing a transverse cross-section are as illustrated in Fig. 5.10b. The total torque on a cross-section is made up part from the torques in the longitudinal members and part from the opposed shear forces on the two sides of the deck. These shear forces are in equilibrium with the torques in the transverse members as shown in Fig. 5.11. The system of forces in Fig. 5.10b is in fact very similar to that in the cellular deck of Fig. 10a. By cutting the slabs between the webs as in Fig. 5.12 we can see that the grillage torques represent the torques in the deck due to the opposed shear flows in the top and bottom slabs, while the grillage shear force S_T represents the shear flow in the webs.

The torsional stiffness of a longitudinal or transverse grillage member is equal to that of the top and bottom slabs represented by the member. Their torsion constant is the same as that of two layers of similar thickness in a solid slab, giving

$$c = 2(h'^2 d' + h''^2 d'') = \frac{2h^2 d' d''}{(d' + d'')} \text{ per unit width of cell.} \tag{5.5}$$

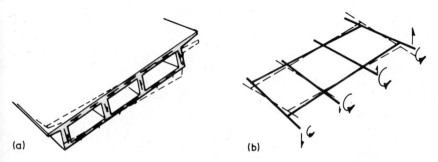

Fig. 5.10 Torsion forces on cross-section of twisted deck and grillage. (a) Shear flows in deck (b) torques and shear forces in grillage.

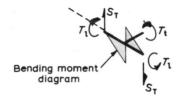

Fig. 5.11 Equilibrium of transverse torque and shear force in edge grillage member.

This constant is equal to half the Saint-Venant torsion constant (see Section 2.4.3) of a wide box beam per unit width. Just as in Chapter 3, the sum of the torsion stiffnesses of the longitudinal members of the grillage is only half the Saint-Venant torsion stiffness of the section if it is treated as a beam. This reflects the fact that when a grillage is twisted the longitudinal member torques are only providing half of the total torque on the cross-section, the other half is provided by the opposed vertical shear forces on opposite sides of the deck. When the deck is twisted by differing amounts across the cross-section, as shown in Fig. 5.13, the relationship between slab and beam torsion constants is no longer relevant. However, the statical equivalence of the grillage and cellular deck force systems still holds.

The webs experience shear deformation due to the shear flows, and hence the longitudinal grillage members should be given shear areas equal to the cross-sectional areas of the webs.

5.4.4 Distortion

Distortion of cells occurs as shown in Fig. 5.3e when the cells have few or no transverse diaphragms or internal bracing, so that a vertical shear force across a

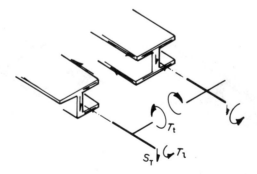

Fig. 5.12 Statically equivalent torsion forces in deck and grillage.

Cellular Decks: Shear-Flexible Grillage Analysis 97

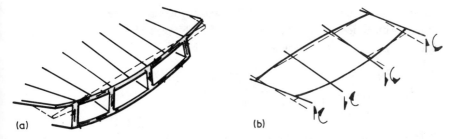

Fig. 5.13 Non-uniform torsion of (a) deck and (b) grillage.

cell causes the slabs and webs to flex independently out-of-plane. This pattern of deformation is very similar to that of a Vierendeel truss of elevation similar to the cross-section of the deck. Although such behaviour cannot be reproduced precisely in a flat grillage, an approximation to this behaviour can be introduced by giving the transverse grillage members a low shear stiffness. The stiffness is chosen so that when the grillage member and cell are subjected to the same shear force, they experience similar distortions as in Fig. 5.14. One error in this analogy is that for the grillage beam the shear force is solely proportional to the shear displacement, while in the cell the shear force is to some extent dependent on the continuity of moments from flexure of the slabs in the adjacent cells. Fortunately it is found that the effect of this difference on the overall structural behaviour is small.

To determine the equivalent shear area of a transverse grillage member we must determine the relationship between vertical shear across a cell and the effective shear displacement w_s shown in Fig. 5.14. Precise frame analysis of a cell with differing thicknesses of top and bottom slabs and webs produces equations with unmanageable complexities. A convenient approximation is obtained by assuming that the shear force is shared between top and bottom

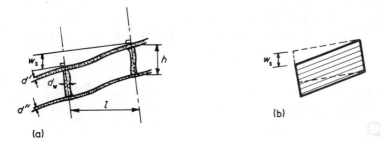

Fig. 5.14 Cell distortion and equivalent shear deformation of grillage member. (a) Cell distortion   (b) shear deformation.

slabs in proportion to their individual flexural stiffnesses and that there are points of contraflexure midway between webs. The vertical shear force per unit width across a cell is then given approximately by

$$s \simeq \frac{(d'^3 + d''^3)}{l^3} \left[\frac{d_w^3 l}{d_w^3 l + (d'^3 + d''^3)h} \right] E w_S \qquad (5.6)$$

where d', d'', d_w, l and h are the dimensions shown in Fig. 5.14.

For the shear flexible grillage member, the relationship between shear force and shear displacement is

$$s = \frac{a_S G w_S}{l} \text{ per unit width of member} \qquad (5.7)$$

where a_S is the equivalent shear area of the member.

Equating the stiffnesses of Equations 5.6 and 5.7 we obtain the following expression for equivalent shear area of the grillage members

$$a_S = \frac{(d'^3 + d''^3)}{l^2} \left[\frac{d_w^3 l}{d_w^3 l + (d'^3 + d''^3)h} \right] \frac{E}{G} \text{ per unit width.} \qquad (5.8)$$

If the webs in the deck are much closer together than it is practicable to place longitudinal grillage members, the transverse a_S is still calculated using the actual cell and web dimensions.

The above expression is strictly only applicable to cells of rectangular cross-section. If a deck has triangular or trapezoidal cells as in Fig. 5.15a, the above equation is not relevant and the shear stiffness must be derived from a frame analysis of a frame of the same shape and dimensions as unit length of the deck. For complicated cross-sections this is most conveniently done using a computer plane frame analysis. The frame is supported as in Fig. 5.15b so that it cannot rotate and subjected to distortional shear forces s. The shear stiffness of each cell is then s divided by the relative vertical movement across the cell. By

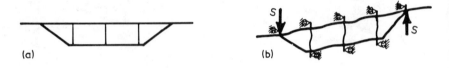

Fig. 5.15 Plane frame analysis of shear stiffness of trapezoidal cells. (a) Deck cross-section   (b) plane frame subjected to distortion.

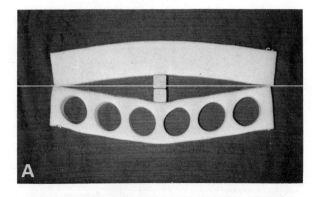
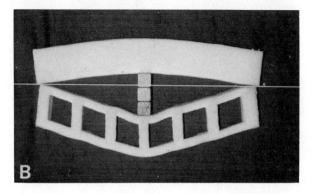

Fig. 5.16 Foam plastic cellular and solid beams flexed back to back to compare deflections under identical loads.

equating the stiffness to $(a_S G/l)$ in Equation 5.7, we obtain the equivalent shear area of the grillage member across the cell.

The transverse shear area of decks with cylindrical voids, as Fig. 5.1d, can be calculated approximately by notionally replacing the cylindrical voids by square section voids of the same cross-section area. Equation 5.8 is then applied using the dimensions of the deck with square voids. Since the square voids are considerably more flexible than the cylinders, the calculated shear stiffness is an underestimate. An alternative method is to make a simple model of the deck cross-section such as the foam plastic model in Fig. 5.16a. The deflections of this model due to distortion can be found by comparing the deflections with those of a solid section beam with the same moment of inertia (as at the centre line of the voids). Fig. 5.16b shows the distortion of a Vierendeel beam with voids of the same area. To minimize the effects of creep, the beams should be loaded and inspected at precisely the same time intervals.

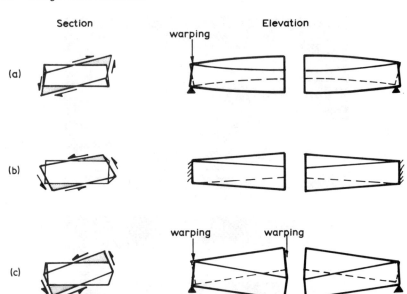

Fig. 5.17 Cell distortion and torsion with in-plane deformation of plates. (a) Distortion with in-plane bending of plates (b) distortion with in-plane shear of plates (c) torsion with in-plane shear of plates.

Where a transverse grillage member represents part of a cell with a diaphragm, the equivalent shear area is much larger and includes the cross-section area of the diaphragm.

Cell distortion is not just resisted by out-of-plane flexure of the top and bottom slabs and the webs but also by in-plane bending and shear of these plate elements. Fig. 5.17a shows the cross-section and elevation of a cell whose faces have experienced in-plane bending, while Fig. 5.17b shows a cell distorted by in-plane shear of the cell. The shear flows that cause the distortion of Fig. 5.17b can be precisely the same as those that produce torsion in Fig. 5.17c. If the section is prevented from distorting, torsion occurs as in Fig. 5.17c without change of cross-section shape but with warping of the cross-section (i.e. alternate longitudinal movements in the corners); if the deck has a higher stiffness against warping than distortion, the shear deformation of the faces causes the cell to distort as in Fig. 5.17b.

Separation of the effects of torsion and distortion can be confusing, and it is much simpler to consider such behaviour as a combination of 'longitudinal torsion' and 'transverse torsion' as shown in Fig. 5.18a and b. In 'longitudinal torsion' of Fig. 5.18a, the top and bottom slabs are being sheared transversely in opposite directions with twisting of the webs but no relative longitudinal slope

Cellular Decks: Shear-Flexible Grillage Analysis   101

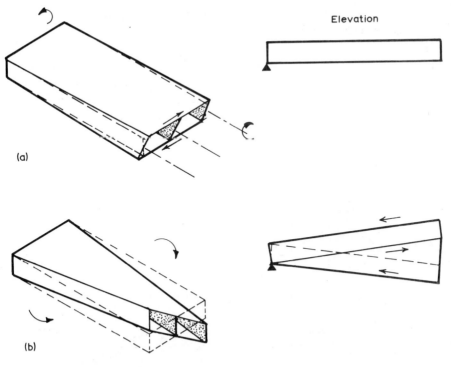

Fig. 5.18   (a) 'Longitudinal torsion'   (b) 'transverse torsion'.

of the webs when viewed in elevation. On the other hand, in 'transverse torsion' the top and bottom slabs are being sheared in opposite directions longitudinally with the webs deflecting with different slopes in elevation but not being twisted. In wide decks the very high in-plane bending stiffnesses of the top and bottom slabs prevents them deflecting laterally in opposite directions so that webs effectively remain vertical and 'transverse torsion' dominates.

While the twisting of a distorted cellular deck is different in the two directions, the transverse and longitudinal shear flows in the top and bottom slabs must still be complementary so that the torques per unit width and per unit length remain equal. This behaviour cannot be precisely represented in a grillage since there is no interaction between longitudinal and transverse torques. However, if the longitudinal and transverse members of the grillage are subjected to the same longitudinal and transverse twists as the cellular structure, the grillage torques can still be statically equivalent to the shear flows in the cellular structure. Fig. 5.19a, b and c show elements of top and bottom slabs of a cellular deck subjected to identical shear flows in pure torsion, 'transverse torsion', and 'longitudinal torsion'. Since the shear flows are identical, so are the shear strains and hence also the sum of the twists in the two directions. In (a) deformation is pure torsion with equal twists $\dot{\phi}_A$ in the two directions; (b) is 'transverse torsion' with transverse twist $\dot{\phi}_t = 2\dot{\phi}_A$ and zero longitudinal twist $\dot{\phi}_l$ and (c) is 'longitudinal torsion' with longitudinal twist $\dot{\phi}_l = 2\dot{\phi}_A$ and $\dot{\phi}_t = 0$. Fig. 5.19a, b and c also shows the grillage members representing this element of cell subjected to the same twists. If their torsion constants are derived with Equation 5.5 for the pure torsion of (a), the torques in (b) and (c) are

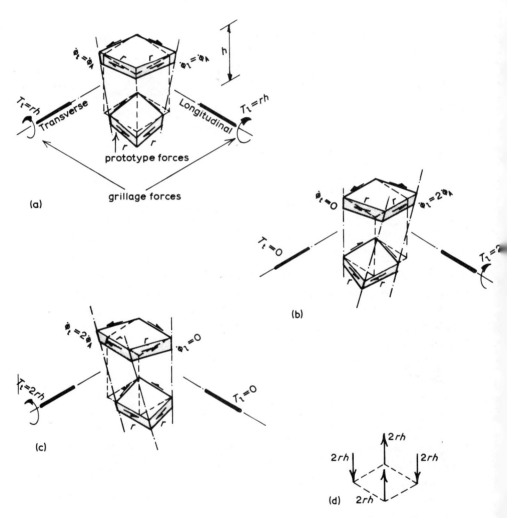

Fig. 5.19 Statical equivalence of torsion forces on element of cell. (a) Pure torsion (b) 'longitudinal torsion' (c) 'transverse torsion' (d) statically equivalent forces.

proportional to the twists, as shown. All these systems of grillage torques and cell shear flows are statically equivalent to the system of forces in Fig. 5.19d and hence to each other. Although the torque in the deck may differ significantly from the grillage torque calculated for a member in one direction, it should be close to the average torque calculated from both transverse and longitudinal members. Hence during the interpretation of the grillage output, the torques per unit width and length of the cell should both be calculated as equal to the average of the torques output for the local transverse and longitudinal members (per unit width of member).

5.5 Section properties of grillage members

This section summarizes the findings of the preceding section and demonstrates calculated examples of grillage section properties for three different decks. The notation for the dimensions of the cells, slabs and webs is defined in Fig. 5.14a.

5.5.1 Grillage for three span twin-cell box girder deck

Fig. 5.20 gives details of a three span twin-cell box girder with supports at 21° skew. The grillage mesh is chosen with three 'structural' longitudinal members 2, 3 and 4 coincident with the webs. Two 'nominal' members 1 and 5 are located along the edges of the cantilevers. Transverse members representing the top and bottom slabs are orthogonal to the longitudinal members. Along the spans their spacing is approximately one quarter of the distance between points of contraflexure, but over the intermediate supports the spacing is shorter to give greater detail near peaks in the bending moment diagrams. At the ends the skew members represent the slabs and the diaphragm, while at the internal supports the skew member represents just the solid diaphragm without flanges.

The 'structural' longitudinal members 2, 3 and 4 have the moments of inertia of the 'I-beams' obtained by cutting the deck as in Fig. 5.20b so that the centroid of each 'I-beam' is on the principal axis of the deck. In this case each beam includes one third of the top slab and one third of the bottom slab. The moment of inertia of each is one third of the total moment of inertia of the deck

$$I_2 = I_3 = I_4 = \frac{1.54}{3} = 0.51 \text{ m}^4.$$

The above calculation has ignored the reduction of effective flange widths of the sections due to shear lag. This is significant in this deck, particularly near the intermediate supports, and a simple correction described in Section 8.2 is strictly necessary.

104 Bridge Deck Behaviour

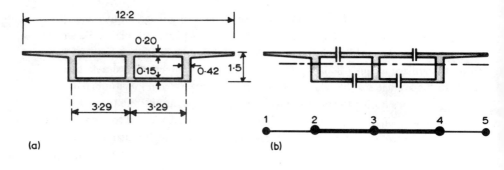

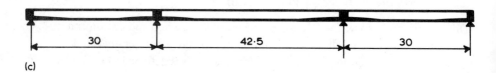

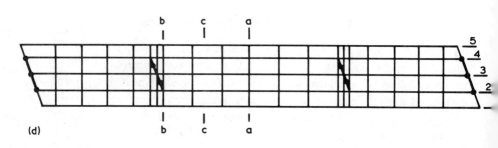

Fig. 5.20 Grillage for three span twin-cell concrete box girder deck. (a) Deck section (b) grillage section (c) deck longitudinal section (d) grillage mesh.

The torsion constant per unit width is given by Equation 5.5

$$c = \frac{2h^2 d' d''}{(d' + d'')} \text{ per unit width}$$

$$c = \frac{2 \times 1.325^2 \times 0.2 \times 0.15}{(0.2 + 0.15)} = 0.30 \text{ m}^4 \text{ m}^{-1}.$$

The widths of cell in members 2, 3 and 4 are $3.29/2$, 3.29 and $3.29/2$ respectively. Hence their torsion constants are

$$C_2 = C_4 = \frac{3.29}{2} \times 0.30 = 0.49 \text{ m}^4 \quad C_3 = 3.29 \times 0.30 = 0.99 \text{ m}^4.$$

The shear areas of members 2, 3 and 4 are equal to the areas of the webs:

$A_{S2} = A_{S3} = A_{S4} = 0.42 \times 1.325 = 0.56 \text{ m}^2$.

Near each support the bottom slab of the deck is thickened. In these regions, the properties of each grillage member are calculated in the same manner as above for the section midway along its length.

The 'nominal' edge members have the section properties of half the cantilever

$$I_1 = I_5 = \frac{bd'^3}{12} = \frac{2.81}{2} \times \frac{0.2^3}{12} = 0.000\,94 \text{ m}^4$$

$$C_1 = C_5 = \frac{bd'^3}{6} = \frac{2.81}{2} \times \frac{0.2^3}{6} = 0.0019 \text{ m}^4$$

$$A_{S1} = A_{S5} = bd' = \frac{2.81}{2} \times 0.2 = 0.28 \text{ m}^2.$$

The transverse members representing cell have section properties given by Equations 5.4, 5.5 and 5.8:

$$i_{23} = \frac{h^2 d' d''}{(d' + d'')} \text{ per unit width} \qquad (5.4)$$

$$= \frac{1.325^2 \times 0.2 \times 0.15}{(0.2 + 0.15)} = 0.15 \text{ m}^4 \text{ m}^{-1}$$

$$c_{23} = \frac{2h^2 d' d''}{(d' + d'')} \text{ per unit width} \qquad (5.5)$$

$$= 2 \times 0.15 = 0.30 \text{ m}^4 \text{ m}^{-1}$$

$$a_{S23} = \frac{(d'^3 + d''^3)}{l^2} \left[\frac{d_w^3 l}{d_w^3 l + (d'^3 + d''^3)h} \right] \frac{E}{G} \text{ per unit width} \qquad (5.8)$$

$$= \frac{(0.2^3 + 0.15^3)}{3.29^2} \left[\frac{0.42^3 \times 3.29}{0.42^3 \times 3.29 + (0.2^3 + 0.15^3)1.325} \right] 2.3$$

$$= 0.0024 \text{ m}^2 \text{m}^{-1}.$$

Transverse members on the cantilever have the properties of the top slab

$$i_{12} = \frac{d^3}{12} \text{ per unit width} = \frac{0.2^3}{12} = 0.000\,67 \text{ m}^4 \text{ m}^{-1}$$

$$c_{12} = \frac{d^3}{6} \text{ per unit width} = \frac{0.2^3}{6} = 0.001\,34 \text{ m}^4 \text{ m}^{-1}$$

$$a_{S12} = d \text{ per unit width} \qquad = 0.2 \text{ m}^2 \text{ m}^{-1}.$$

106 Bridge Deck Behaviour

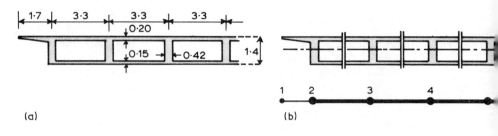

(a)

(b)

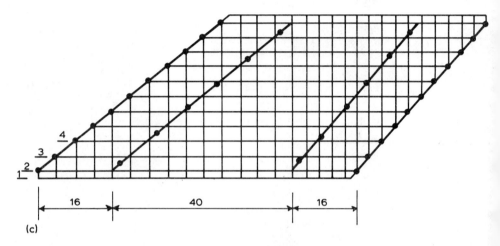

(c)

Fig. 5.21 Grillage for three span high-skew multicell concrete deck. (a) Part deck section (b) grillage section (c) grillage mesh.

The skew members representing the internal diaphragm (here 1.5 m wide) have the properties of the solid section. For derivation of the torsion constant see Section 2.4.2.

$$I = \frac{1.5 \times 1.325^3}{12} = 0.29 \text{ m}^4$$

$$C = \frac{3 \times 1.5^3 \times 1.325^3}{10(1.5^2 + 1.325^2)} = 0.59 \text{ m}^4$$

$A_S = 1.5 \times 1.325 = 2.0 \text{ m}^2$.

5.5.2 Grillage for wide multicellular deck

Fig. 5.21 gives details of a three span multicellular deck at high skew angle. The section properties of the grillage members are derived as the preceeding example except that it is inconvenient to split the deck up into beams with individual centroids precisely on the principal axis of the deck. Consequently, the deck is notionally cut midway between webs as shown in (b). The centroids of internal 'structural' members 3, 4, etc. are virtually coincident with the principal axis of the deck. Edge 'structural' member 2 has its centroid at a higher level, but like the other members its grillage moment of inertia is calculated about the principal axis of the deck.

5.5.3 Grillage for cellular deck with inclined webs

Fig. 5.22 gives details of part of a multispan four-cell deck with inclined edge webs and haunches at the supports. It is impracticable to split up the deck into

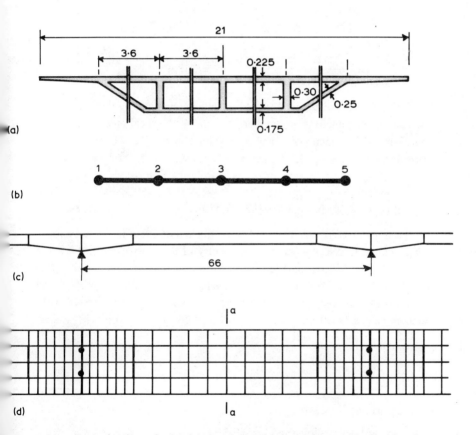

Fig. 5.22 Part of grillage for multispan deck with haunches and inclined webs. (a) Deck section (b) grillage section (c) deck longitudinal section (d) grillage mesh.

longitudinal members with centroids on the principal axis of the deck, hence the deck is notionally cut as in (a) into five 'structural' members with inertias calculated about the principal axis of the deck. There are no 'nominal' edge members in order to permit economy in the size of the grillage.

There is no clear cut equivalence of grillage torsion stiffness for the non-rectangular cells. However, sensible results are obtained if Equation 5.6 is used with h equal to the average height of the cell. Hence

$$C_1 = C_5 = \frac{3.6}{2} \times \frac{2 \times 0.5^2 \times 0.225 \times 0.25}{(0.225 + 0.25)} = 0.1 \text{ m}^4$$

$$C_2 = C_4 = 0.9 \times \frac{2 \times 1.35^2 \times 0.225 \times 0.25}{(0.225 + 0.25)}$$

$$+ 2.7 \times \frac{2 \times 1.7^2 \times 0.225 \times 0.175}{(0.225 + 0.175)} = 1.9 \text{ m}^4$$

$$C_3 = 3.6 \times \frac{2 \times 1.7^2 \times 0.225 \times 0.175}{(0.225 + 0.175)} = 2.0 \text{ m}^4.$$

The shear areas of the longitudinal members must also be derived with engineering judgement. For members 2, 3 and 4, the shear area is the area of the web. For the edge members 1 and 5, the area is somewhat arbitrary but not critical to the analysis, and the figure below is suggested.

$$A_{S1} = A_{S5} = 0.9 \times 0.25 = 0.21 \text{ m}^2$$

$$A_{S2} = A_{S3} = A_{S4} = 1.7 \times 0.3 = 0.51 \text{ m}^2.$$

The transverse grillage members also require special consideration and judgement related to the particular geometry of the cross-section. It is suggested that the moment of inertia and torsion constant can generally be calculated with Equations 5.4 and 5.5 using the average value of h across the cell. The shear areas must be derived from a plane frame analysis as described in Section 5.4.4. Such an analysis of the cross-section of Fig. 5.26a gave the following values:

edge cells $\quad A_{S12} = 0.05 \text{ m}^2 \text{ m}^{-1}$

internal cells $\quad A_{S23} = 0.005 \text{ m}^2 \text{ m}^{-1}.$

5.6 Load application

The webs of many cellular and box girder decks are spaced further apart than the width of the carriageway lanes. Consequently it is possible for a whole lane of loading to lie between grillage members. These loads can be applied to the grillage joints on each side by statical distribution. Since the deck has high transverse bending and longitudinal torsional stiffnesses, its behaviour under these statically distributed loads is virtually the same as if the loads had been applied more correctly as the fixed edge shear forces and moments at the edges of the top slab. This contrasts with the behaviour of a spaced beam-and-slab deck described in Section 4.6 where it was shown that with its low transverse bending and longitudinal torsional stiffnesses, the beam-and-slab deck deflects in different ways when a load is placed between and when it is statically distributed onto the beams.

The grillage of a cellular deck, like that of a beam-and-slab deck, only gives the force systems in the deck due to deformation of the structure as a whole. It does not give an indication of the local moments and shear forces due to the concentration of load between grillage members. These moments and shear forces must be derived independently using the charts of Pucher [2] as described for beam-and-slab decks in Section 4.6 and added to those from the load distribution in the grillage.

5.7 Interpretation of output

The output of the grillage of a cellular deck should be interpretated with as much care as the calculation of section properties. A considerable amount of valuable detailed information about the forces in the cells can be derived from the grillage by reapplying the principle of local static equivalence of forces in cellular deck and grillage.

The following discussion is illustrated with examples of stresses, etc. calculated from the grillage analysis of the deck in Fig. 5.20. The load case for which they are relevant comprised full lane loading and footpath loading on one side of the main span.

5.7.1 Longitudinal bending

The bending moment diagrams for the three 'structural' longitudinal grillage members are shown in Fig. 5.23. The diagrams have saw teeth with large jumps in moment at the joints because of the transfer of the torsion in the transverse members at each joint to bending moments and shear forces in the longitudinal member as was shown in Fig. 5.11. The 'true' bending moment diagrams can be assumed to pass through the average values of the bending moment on the two sides of each joint as shown by the dashed lines in Fig. 5.23.

The longitudinal bending stresses at a section are calculated from these 'true' moments using the section properties of the I-beam represented by the grillage

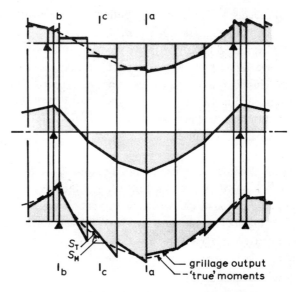

Fig. 5.23   Longitudinal moments in part of grillage.

member. Fig. 5.24a and b show the bending stresses calculated in this way for sections a–a and b–b of the deck. The bending stresses are shown constant across each I-beam without smoothing out the impossible jumps in stress at the notional cuts between I-beams and without correction for shear lag as described in Chapter 8.

The shear force S_M due to bending is the slope of the 'true' dashed bending

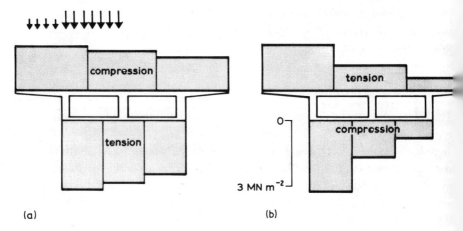

Fig. 5.24   Longitudinal bending stresses. (a) Sagging at midspan a–a (b) hogging at support b–b.

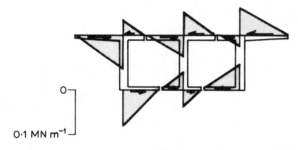

Fig. 5.25 Longitudinal bending shear flow at section c–c.

moment diagram in Fig. 5.23. The shear flows in the web and flanges of the 'I-beams' are calculated from S_M using Equation 5.2. Fig. 5.25 shows these bending shear flows at cross-section c–c.

5.7.2 Transverse bending

The transverse bending moment in the grillage member is equivalent to the opposed transverse compression of the top slab and tension of the bottom slab (or vice versa) due to transverse flexure without distortion. In narrow decks it is usually very small compared to the longitudinal bending moment (except in the diaphragms). However, in wide decks it can be large especially near skew supports. Since these moments interact with the torsions in longitudinal grillage members, a grillage output transverse moment diagram has a saw-tooth shape like the longitudinal moment diagram. And in a similar way the top and bottom slab stresses are calculated from the average moments on the two sides of each joint.

5.7.3 Slab flexure from distortion

The stiffness of the cell against independent flexure of the slabs during distortion is represented by the shear stiffness of transverse grillage members. For this reason the slab bending moments are derived from the shear force in the transverse grillage members. Fig. 5.26a shows the shear force in the transverse members on section a–a of Fig. 5.20d. The fractions of this shear force carried by each of the top and bottom slabs are assumed to be proportional to the flexural stiffnesses of the slabs (here 0.7 : 0.3). Assuming also that points of contraflexure lie midway between the webs, the moment at each end of a slab is simply the shear force it carries times half the distance between webs. Hence we obtain the transverse moment diagram of Fig. 5.26b from the shear forces of

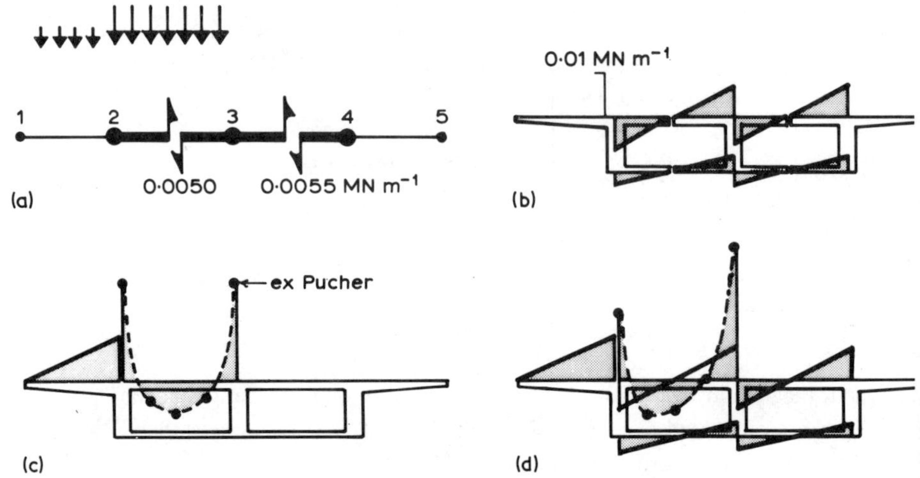

Fig. 5.26 Slab transverse moments at section a—a. (a) Grillage transverse shear forces (b) cell distortion moments (c) cantilever and local moments (d) total moments.

(a). The transverse slab moment in the cantilever can be taken direct from the grillage output since this member is not representing a cell. Fig. 5.26c shows the cantilever moment and the local moments under the knife edge load above section a—a, while (d) shows the total moments obtained by adding (b) and (c). The local moments were derived using the influence charts of Pucher [2] as described in Section 4.6.

5.7.4 Torsion shear flow

The torsion shear flows in the slabs must be calculated from the average torque per unit width of transverse and longitudinal grillage member as described in Section 5.4.4. Fig. 5.27a shows the torques in grillage members per unit width of cell; the averages of the values in the two directions are also shown. By dividing these average torques per unit width of cell by the distance h between slab midplanes, we obtain the shear flows in the slabs shown in Fig. 5.27b. On adding these to the bending shear flows in Fig. 5.25, we obtain the total shear flows shown in Fig. 5.28.

The shear forces output from the grillage for longitudinal members are the slopes of the saw teeth of the output moment diagram of Fig. 5.23. These shear forces combine the components S_M due to bending (the slope of the dashed 'true' moment diagram) and the components S_T due to torsion (the additional slope of the teeth caused by transverse torsion). Consequently, the grillage output shear force represents the total shear force in each web of the deck.

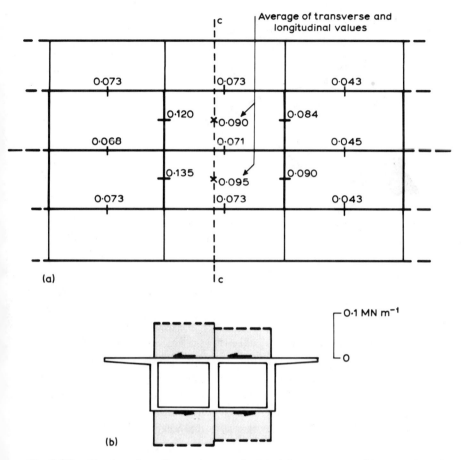

Fig. 5.27 Torsion shear flows at c–c calculated from average of transverse and longitudinal torques output by grillage. (a) Torques in grillage members per unit width (b) shear flows.

Since the torsional shear force S_T in longitudinal members results from equilibrium with torques in transverse members, it is in error if the transverse torque differs significantly from the average of the torques in the two directions. Although it is seldom necessary, a correction can be made to the torsional shear force by decreasing it or increasing it as appropriate in proportion to the excess of the transverse torque over the average torque.

5.8 Other methods of analysis

While the shear-flexible grillage is the most convenient method of analysis for cellular decks, it is not the only method available. Consequently, it is worth

114 *Bridge Deck Behaviour*

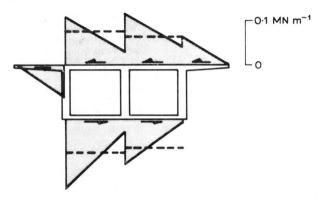

Fig. 5.28   Combined shear flow from torsion and bending.

mentioning some of the other methods in case one is more appropriate for a particular case. A comprehensive review of current methods is provided by Maisel [3].

Continuous beam analysis (described in Chapter 2) is the simplest method of investigating bending moments, torsions, etc. in narrow, right decks (with box width less than about 1/10 of the effective span). It can also be extremely useful when used in conjunction with a grillage for the analysis of right multispan viaducts with wide decks. The continuous beam analysis is used to investigate the distribution along the spans of total moments, etc. on cross-sections due to dead load, construction sequence, prestress, and various spanwise distributions of live loading. The grillage is then used for three or four typical spans to determine the distributions across the deck of moment, etc. and to study effects of various transverse dispositions of lane loading. An alternative to this application of grillage analysis is sometimes possible using the beam on elastic foundations analogy [4]. This method, however, is neither as convenient nor as versatile as the grillage.

Cellular decks of complex or unusual construction generally require some form of analysis in addition to grillage to investigate the plate behaviour of the elements and stress concentrations near diaphragms. Decks with prismatic form (same section from end to end), right end supports, and few diaphragms can usually be analysed relatively easily with the folded plate and finite strip methods described in Chapters 12 and 13. If the deck has varying section or skew supports or numerous diaphragms, a space frame analysis can be carried out without too much complexity. The frame members are arranged in lattice or cruciform mesh as described in Chapter 7. Very complicated decks will usually require investigation with a finite element analysis as described in Chapter 13. Because of the complexity and expense of this method, it is often more convenient and economic to use a grillage for an approximate analysis of load distribution and effects of construction procedure and then use a finite element analysis to investigate small regions subject to critical and complex stresses.

REFERENCES

1. Sawko, F. (1968), 'Recent developments in the analysis of steel bridges using electronic computers,' BCSA Conference on Steel Bridges, London, pp. 39–48.
2. Pucher, A. (1964), *Influence Surfaces of Elastic Plates,* Springer Verlag, Wien and New York.
3. Maisel, B. I. (1970), 'Review of literature related to the analysis and design of thin walled beams,' Technical Report 42.440, Cement and Concrete Association, London, July.
4. Wright, R. N., Abdel-Samad, S. R. and Robinson, A. R., (1968), 'BEF for analysis of box girders,' *Journal of the Structural Division of the A.S.C.E.,* **94** ST7, 1719–1743.

6
Shear key decks

6.1 Introduction

A shear key deck, shown in Fig. 6.1, is constructed of a number of parallel contiguous beams attached to each other along their length by stitch joints which have low transverse bending stiffness. There is no intrinsic difference between shear key decks and beam-and-slab decks, described in Chapter 4, but the particular structure and stiffnesses of the shear key deck give it particular characteristics. Prior to the development of the computer aided grillage, such decks were generally analysed with the 'articulated plate theory' developed by Spindel [1]. This theory is the basis of the load distribution coefficient charts of reference [2]. However, the most convenient method at present is a special application of the grillage discussed in Chapters 3 and 4. This chapter describes the characteristics of the shear key deck and the appropriate details of its simulation by grillage.

6.2 Structural behaviour

The longitudinal joints between the beams can be thought of as full length 'piano' hinges. When the deck is loaded as in Fig. 6.2, part of the load is carried by the beams beneath and part transferred laterally to neighbouring beams by vertical shear forces on the hinges. Unlike the beam-and-slab decks of Chapter 4,

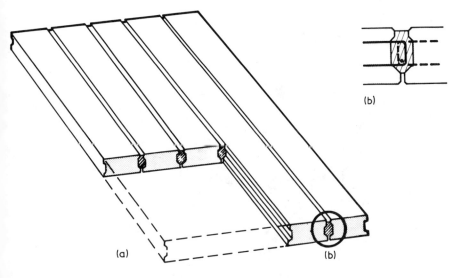

Fig. 6.1   Shear key deck.

this transverse shear is resisted primarily by the torsional stiffness of the beams and only to a small extent by the transverse bending stiffness of the articulated structure (which is nil if the joints are true hinges).

The forces acting on an element of beam, shown in Fig. 6.3, are similar to those for beam-and-slab deck in Fig. 4.3. Hence equilibrium equations are the same as Equation 4.1.

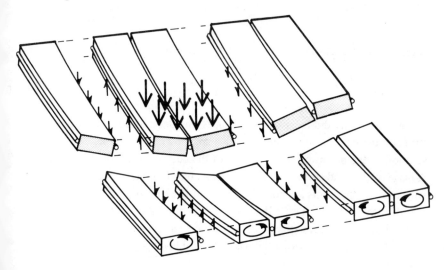

Fig. 6.2   Transverse load distribution by vertical shear resisted by beam torsion.

118 Bridge Deck Behaviour

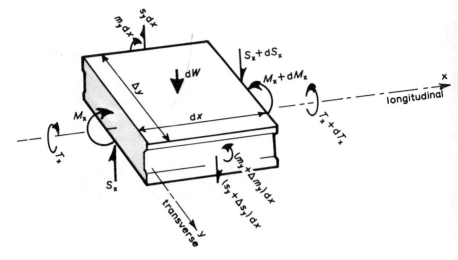

Fig. 6.3   Element of beam of shear key deck.

$$\frac{dS_x}{dx} + \Delta s_y = -W\Delta y$$

$$\frac{dM_x}{dx} = S_x \tag{4.1}$$

$$\Delta m_y + \frac{dT_x}{dx} = s_y \Delta y.$$

When the flexural stiffness of the joints is negligible the third equation reduces to

$$\frac{dT_x}{dx} = s_y \Delta y. \tag{6.1}$$

If the joints really did behave as free hinges, the deck could be expected to deflect under a point load with a sharp cusp, shown in Fig. 6.4a. The relative rotation of the beams on the two sides of the cusp is then so large that crushing damage of the road surfacing might be expected. In practice, the flexural stiffnesses of the joints are not negligible and the behaviour of the deck is significantly modified. Best [3] showed that with relatively little reinforcement in the shear keys, the relative rotation of the beams is considerably reduced below the levels predicted on the basis of flexible hinges, as illustrated in Fig. 6.4b.

Since in practice the bending stiffness of some joint configurations is difficult

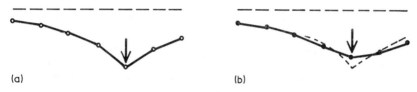

Fig. 6.4 Midspan deflections of shear key deck supporting point or line load.
(a) Free joints (b) partially stiff joints.

to predict, it is often sensible to assume for purposes of design that the joints are flexible. This results in conservative estimates for maximum beam torsions. Nonetheless, unless the deck is somehow constructed with the real hinges, the ignored joint stiffness prevents damaging transverse crushing strains occurring in the surfacing.

Shear key decks can be stiffened at the edges with upstand beams as described for slabs in Section 8.3. If a width of the slab is considered as flange to the upstand, a check should be made that the shear keys can transmit the longitudinal shear flow into the flange.

6.3 Grillage analysis

6.3.1 Grillage structure

A shear key deck such as in Fig. 6.5a can be represented by the grillage of (b) which has longitudinal members coincident with the centre lines of the beams of the prototype. Each longitudinal grillage member has transverse outriggers which are stiff. The shear keys are represented by the pinned joints between outriggers of adjacent beams.

If the shear keys do have bending stiffness (or if the computer program cannot accommodate pinned joints), the outriggers can be connected by short flexible members with all joints stiff as in (c). The length and stiffness of the flexible connection should simulate as accurately as possible the dimensions and stiffness of the shear key. For example, it is possible for the shear key to consist simply of steel dowels; in this case, the short flexible grillage members are given the effective length and stiffness of the dowels. One pitfall that can be encountered if a grillage has members with stiffnesses differing by several orders of magnitude is that the computer may not calculate with a sufficient number of significant figures so that solution of the stiffness equations may prove inaccurate or impossible. It is worth checking beforehand that the computer is sufficiently accurate.

The grillages of (b) and (c) have very large numbers of members with the result

120 *Bridge Deck Behaviour*

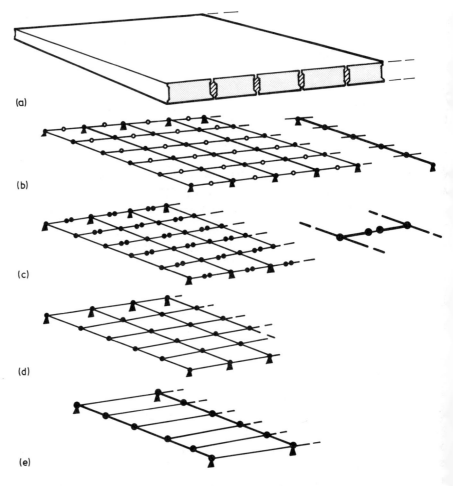

Fig. 6.5   Grillage representations of shear key deck.

that the quantity of computer data and output is unmanageable. The number of members can be dramatically reduced in two ways. Firstly, the chains of transverse members of varying stiffness can be replaced by single members of equivalent stiffness as illustrated in (d) and explained below. Secondly, two or more deck beams can be represented by each longitudinal grillage member as in (e). The grillage members can be placed with lateral spacing of up to 1/5 span without seriously affecting the magnitude of calculated peak moments under a lane load. Such expediency might appear to reduce the accuracy of the analysis, but if it is done with care there is little discernable difference in the result, and the risk of arithmetic error is considerably reduced. Furthermore, when the

Shear Key Decks

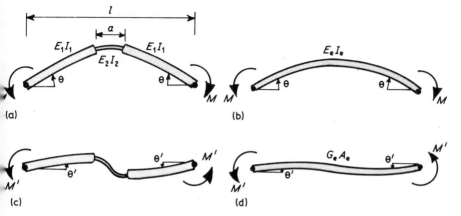

Fig. 6.6 Representation of chain of members with different inertias by equivalent uniform grillage member. (a and b) Symmetric bending (c and d) antisymmetric bending.

overall analytical approximations are compared with the structure as built, it is often difficult to say which analysis is closest to the very different reality.

The chain of transverse members of various stiffnesses in Fig. 6.6a has the same stiffness under symmetric bending as the equivalent member of (b) if

$$\frac{M}{\theta} = \frac{2}{\int \frac{dx}{EI}} = \frac{2}{\frac{l-a}{E_1 I_1} + \frac{a}{E_2 I_2}} = \frac{2}{\frac{l}{E_e I_e}} \tag{6.2}$$

where $E_1 I_2$, $E_2 I_2$ are the flexural rigidities of the segments of the variable stiffness prototype, and $E_e I_e$ is the flexural rigidity of the equivalent grillage member.

Under the antisymmetric bending of (c) the equivalent member of (d) will have a different equivalent stiffness from that given by Equation 6.2 unless shear flexibility of the equivalent member is considered. For the beams of varying inertia in (c) the stiffness is given by:

$$\frac{M'}{\theta'} = \frac{l^2}{2I_0} \tag{6.3}$$

$I_0 = \int \frac{x^2 dx}{EI}$ = the second moment of area of $\frac{dx}{EI}$ elements about their centroid.

$$I_0 = \frac{(l^3 - a^3)}{12 E_1 I_1} + \frac{a^3}{12 E_2 I_2}.$$

It should be noted that these equations assume the beams of different inertia are symmetrically disposed about the midpoint.

For the equivalent shear flexible beam of (d) the stiffness is

$$\frac{M'}{\theta'} = \frac{l^2}{2\left(\dfrac{l^3}{12E_e I_e} + \dfrac{l}{G_e A_e}\right)} \qquad (6.4)$$

where $A_e G_e$ is the shear rigidity of the equivalent grillage member.

Whence we obtain for equivalence

$$\frac{l^2}{2\left(\dfrac{l^3 - a^3}{12E_1 I_1} + \dfrac{a^3}{12E_2 I_2}\right)} = \frac{l^2}{2\left(\dfrac{l^3}{12E_e I_e} + \dfrac{l}{G_e A_e}\right)} \qquad (6.5)$$

Consequently, from Equations 6.2 and 6.5 the section properties of the equivalent member are

$$\frac{1}{I_e} = \left(1 - \frac{a}{l}\right)\frac{E_e}{E_1 I_1} + \frac{a}{l}\frac{E_e}{E_2 I_2}$$

$$\frac{1}{A_e} = \left[\left(1 - \frac{a^3}{l^3}\right)\frac{1}{E_1 I_1} + \frac{a^3}{l^3}\frac{1}{E_2 I_2} - \frac{1}{E_e I_e}\right]\frac{l^2 G_e}{12} \qquad (6.6)$$

It will be found that A_e is negative.

An exception to the above equivalence arises when the shear key has such a low flexural stiffness that it is effectively pin jointed. In this case Equation 6.6 is indeterminate, and a real pinned joint should be included in the grillage member.

Shear key decks can be constructed of thin-walled box beams whose cross-sections distort under the action of the transverse shear force. Such distortion can be included in the grillage analysis by suitably modifying the equivalent shear area of transverse members as described for cellular decks in Section 5.4.4. The simplest method of determining the equivalent shear area of the complicated section is to carry out a computer analysis of a plane frame of shape similar to the deck cross-section as demonstrated for a cellular deck in Fig. 5.15.

Shear key deck theory and charts can also be used for the analysis of spaced box beam-and-slab decks such as in Fig. 4.1. The torsional stiffness of the boxes makes them behave in a manner similar to the shear key beams, with the thin slab behaving like a very flexible shear key.

6.3.2 Example of section properties

Fig. 6.7a shows the cross-section of the beams represented by each longitudinal grillage member in Fig. 6.5e. The moment of inertia of the longitudinal member equals the sum of the inertias of the beams it represents. Here

$$I_x = 2 \times \frac{0.902 \times 0.381^3}{12} = 0.0083.$$

Since the joints do not provide continuity of torsion shear flow from one beam to the next, the torsion shear flow in each is a closed loop as in Fig. 6.8. This torsional behaviour is identical to that for an isolated beam for which the torsion constant is given in Section 2.4.3. The torsion constant of the grillage member is simply the sum of the Saint-Venant torsion constants of the constituent beams. Here

$$C_x = 2 \times \frac{3 \times 0.902^3 \times 0.381^3}{10(0.902^2 + 0.381^2)} = 0.025.$$

It should be noted that this torsional behaviour is totally different from that of the orthotropic slab deck of Fig. 3.1d in which transverse prestress enables torsional shear flow interaction between beams. Then the torsion shear flow on the gross cross-section flows round the whole section as in Fig. 3.5.

The transverse grillage member must be equivalent to the varying transverse section of Fig. 6.7b. The inertia of the joint is calculated for the transformed cracked section of reinforced concrete under sagging moment, the bottom transverse reinforcement being considered as the steel in tension. The inertia of

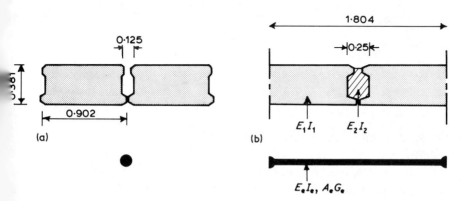

Fig. 6.7 Details for grillage of Fig. 6.5e. (a) Section of longitudinal member (b) transverse member.

Fig. 6.8   Torsion shear flow in beams of shear key deck.

the beams on each side is calculated as for uncracked concrete, with reinforcement ignored. Using Equation 6.6, with properties per unit length

$$E_2 = 0.8E_1, \quad E_e = E_1, \quad G_e = 0.43E_1, \quad l = 1.804 \quad a = 0.25$$

$$i_1 = \frac{0.381^3}{12} = 4.6 \times 10^{-3} \quad i_2 = 4.1 \times 10^{-4}$$

$$i_e = 1 \bigg/ \left(\frac{0.86}{4.6 \times 10^{-3}} + \frac{0.14}{0.8 \times 4.1 \times 10^{-4}} \right) = 1.65 \times 10^{-3}$$

$$a_e = 1 \bigg/ \left(\frac{0.997}{4.6 \times 10^{-3}} + \frac{0.0027}{0.8 \times 4.1 \times 10^{-4}} - \frac{1}{1.65 \times 10^{-3}} \right) \frac{1.804^2 \times 0.43}{12}$$

$$= -0.022.$$

The joint has little transverse torsional stiffness and a nominal torsion constant can be attributed to transverse grillage members. Alternatively, a value can be obtained using the equation $c = l d^3/6$ appropriate to a slab with d equal to depth of compression concrete calculated during computation of the joint's moment of inertia. Here

$$c = \frac{0.053^3}{6} = 2.5 \times 10^{-5}.$$

6.4 Skew decks

Shear key decks can be constructed at high skew angles. The load distribution differs from that in right decks in that longitudinal bending moments are smaller while beam torques are considerably larger. If the shear keys have little or no transverse bending stiffness, the beam torques may be larger than the design strength that can be given to them. Consequently it is essential to provide some moment stiffness to the shear keys and also essential to make a realistic estimate of this stiffness in the analysis.

The supports of skew shear key decks, as of other types of skew deck, should not be too stiff. Not only does such stiffness increase the beam torsions but it will result in some bearings being subjected to high reactions while others near the acute corner will be subject to uplift. In addition, if the bearings of the prototype are very stiff, the distribution of reactions and beam end shear forces is unpredictable [2], with local variation in joint stiffness having a significant effect. In contrast, if the bearings are soft, the distribution of reactions is more uniform and predictable.

REFERENCES

1. Spindel, J. E. (1961), 'A study of bridge slabs having no transverse flexural stiffness,' PhD Thesis, Kings College, London.
2. Department of Environment (1970), 'Ministry of Transport Technical Memorandum Shear Key Decks,' Annexe to Technical Memorandum (Bridges) No. BE 23.
3. Best, B. C. (1963), 'Tests of a prestressed concrete bridge incorporating transverse mild-steel shear connectors,' Cement and Concrete Association, Research Report 16.

7
Three-dimensional space frame analyses and slab membrane action

7.1 Slab membrane action

It was shown in Section 4.8 that the conventional plane grillage does not reproduce transfer of in-plane shear across the slab strips between the longitudinal beams of a beam-and-slab deck. Fig. 7.1a shows such a slab strip subjected to in-plane shear flows $r_{1\,2}$ and $r_{2\,1}$ along its edge. These shear flows can be split into the symmetric shear flows in (b) and antisymmetric shear flows in (c). The conventional grillage analysis simulates the shear flows of (b) by considering the two halves of the slab strip as flanges of the T-beams to each side. It is the antisymmetric inter-beam shear flow of (c) that is ignored.

The antisymmetric shear flows of Fig. 7.1c can cause two types of in-plane deformation depending on whether they are accompanied by transverse compression forces along the edges. Fig. 7.2a shows a narrow slab strip subjected to antisymmetric shear flows in the absence of transverse forces, and the strip deflects by in-plane bending with cross-sections fanning radially. In contrast, (b) shows a thin slab strip distorting in-plane by shear deformation with cross-sections remaining parallel and rectangular elements shearing to parallelograms. An isolated narrow strip is very much stiffer in shear than in bending, so that if the slab represents the back of a narrow channel under antisymmetric loads as in Fig. 7.3, the edge shear flows cause it to deflect sideways in bending (in direction towards the stretched edge). However, if the

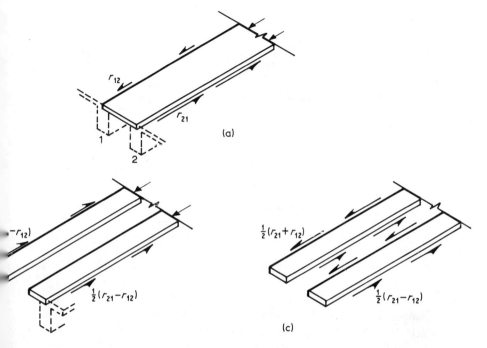

Fig. 7.1 Shear flow in slab between beams of beam-and-slab deck. (a) Total shear flow (b) flange shear flow (c) inter-beam shear flow.

strip is stiffened against sideways displacement by numerous other strips forming a wide slab, the edges remain virtually straight and the slab distorts in 'trapezoidal shear' as in Fig. 7.4. This is a combination of the in-plane bending of Fig. 7.2a and in-plane shear of Fig. 7.2b. Because the strip has much greater stiffness for in-plane shear than for in-plane bending, the dominating forces and stresses associated with such distortion relate to shear.

7.2 Downstand grillage

The in-plane inter-beam shear of the slab can be incorporated in any of several three-dimensional methods including folded plate theory described in

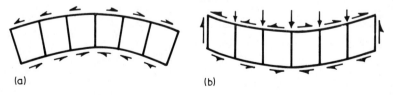

Fig. 7.2 In-plane deformation of strip of slab. (a) Bending (b) shear.

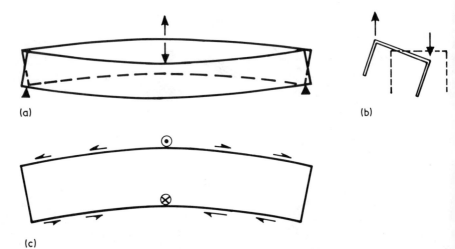

Fig. 7.3 Sideways flexure of channel. (a) Elevation (b) section (c) plan of top slab.

Chapter 12 and finite elements in Chapter 13. The cheapest computer-aided methods at present are special applications of space frame analysis. These are described in the remainder of this chapter.

7.2.1 Downstand grillage

Most road bridges of beam-and-slab construction can be analysed as three-dimensional structures by a space frame analysis which is a simple extension of the grillage described in Chapter 4. Instead of a deck such as in Fig. 7.5a being represented by a plane grillage in (b), a space frame is used, as in (c). The mesh of the space frame in plan is identical to the grillage, but the various transverse and longitudinal members are placed coincident with the line of the centroids of the downstand or upstand member they represent. For this

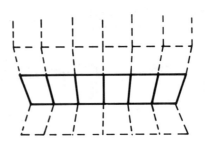

Fig. 7.4   'Trapezoidal shear' of strip of slab.

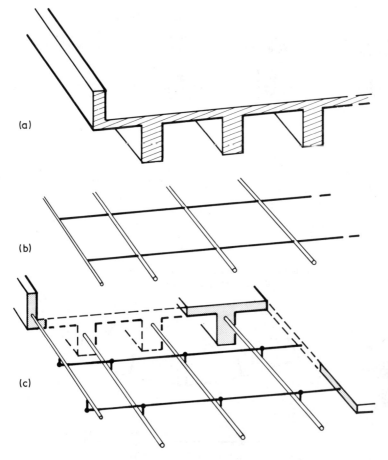

Fig. 7.5 Downstand grillage representation of beam-and-slab deck. (a) Deck (b) plane grillage (c) downstand grillage.

reason the space frame is here referred to as a 'downstand grillage'. The longitudinal and transverse members are joined by vertical members which, being short, are very stiff in bending.

The downstand grillage behaves in a similar fashion to the plane grillage under actions of transverse and longitudinal torsion and bending in a vertical plane. Consequently, the section properties for these actions are calculated in the same way. Thus for the deck of Fig. 7.6 we have, as in Section 4.5.2,

$$I_x = 0.21 \qquad C_x = 0.0032$$

$$I_y = \frac{4 \times 0.2^3}{12} = 0.0027 \qquad C_y = \frac{4 \times 0.2^3}{6} = 0.0053.$$

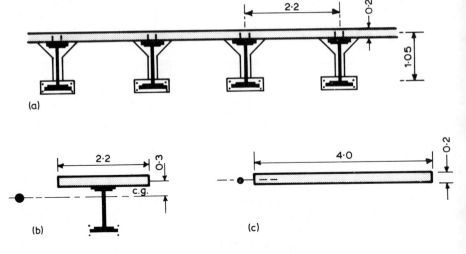

Fig. 7.6 Dimensions of composite steel and concrete deck. (a) Section of deck (b) longitudinal member (c) transverse member.

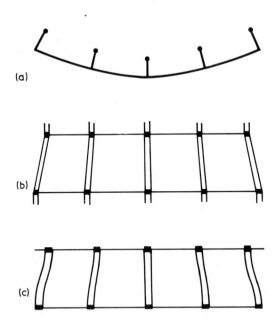

Fig. 7.7 In-plane shear of transverse members of downstand grillage representing slab. (a) Warping displacement due to bending of longitudinal members (b) plan view of in-plane shear of transverse members (c) erroneous in-plane flexure of transverse members.

Vertical deflection of the downstand beams causes joints in plane of slab to move longitudinally as in Fig. 7.7a. These warping displacements generate in-plane shear forces in the transverse members if the members are given the shear areas of the cross-section of slab they represent and if they are forced to shear as in Fig. 7.7b. Curvilinear distortion of the mesh as in Fig. 7.7c must be prevented by restraining all slab joints against sideways displacement v and rotation θ_z about vertical axis. In addition, transverse members are given very high inertias for in-plane bending so that shear deformation dominates. Thus for in-plane shear and moment in the slab of Fig. 7.6,

$$A = bd = 4.0 \times 0.2 = 0.8$$

$$I = \text{say}, \quad \frac{(\text{deck span})^3 \times d}{12} = 800$$

Joints in slab $v = \theta_z = 0$.

It should be noted that because the joints in the slab are restrained against transverse horizontal displacement, the downstand grillage model is not appropriate for analysis of sections, like the U-beam in Fig. 7.3, which have deep downstands and overall width less than about 1/3 effective span. The transverse flexure of such decks is significant. If there is doubt about lateral bending of the deck, McHenry or cruciform lattices described in Sections 7.3 and 7.4 should be used.

The transfer of in-plane shear by transverse members subjects the longitudinal members to axial loads, as shown in Fig. 4.13b. The longitudinal members must be able to stretch (or compress) under these axial forces and so each is given the cross-sectional area of the part of deck it represents, including slab flanges. For the deck of Fig. 7.6, with modular ratio $m = 7$ for steel beam

$$A_x = \text{longitudinal T-beam area} = 0.09 \times 7 + 2.2 \times 0.2 = 1.27.$$

If the longitudinal beams are spaced at more than 1/6 effective span, shear lag reduces the effective width of flange and inter-beam shear transfer is small. To minimize error, the width of slab associated with any downstand beam should be restricted to less than 1/12 effective span. Any slab between downstands and not included in their flanges is considered as a separate longitudinal beam. The only difference between it and downstand longitudinal members is that the vertical downstand members are non-existent.

Under the action of transverse member bending and longitudinal member torsion, the longitudinal members deflect sideways as in Fig. 7.8. The bottom flange contributes most of the flexural stiffness of the beam against such

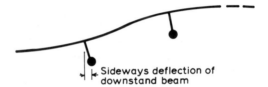

Fig. 7.8 Part section of downstand grillage showing sideways deflection of longitudinal beam due to transverse bending.

sideways deflection. Since the longitudinal member at the centroid deflects sideways less than bottom flange in ratio z_c/z_b of their respective distances from the top slab, the member should be given an effective inertia z_b/z_c times that of bottom flange. Hence for sideways deflection of beam

$$I = \frac{1.05 \times 0.000\,68}{0.30} \times 7 = 0.017.$$

The vertical members behave in similar manner to transverse slab members except that plane of action is rotated through 90°. Hence for these in Fig. 7.6,

out-of-plane

$$I = \frac{bd^3}{12} = \frac{4 \times 0.021^3}{12} \times 7 = 2.2 \times 10^{-5}$$

$$C = \frac{bd^3}{6} = \frac{4 \times 0.021^3}{6} \times 7 = 4.4 \times 10^{-5}$$

$$A = bd = 4.0 \times 0.021 \times 7 = 0.59.$$

in-plane

$$I = \text{say} \quad \frac{(\text{span})^3 d}{12} = 600$$

$$A = bd = 4.0 \times 0.021 \times 7 = 0.59.$$

7.2.2 Interpretation of downstand grillage output

Fig. 7.9 shows the element of beam of Fig. 4.3 but with inter-beam shear flows r_{01} and r_{12} and balancing axial tension P. The equilibrium of the element differs in two ways from Equations 4.1. Firstly on resolving longitudinally,

$$\frac{dP_x}{dx} + (r_{12} - r_{01}) = 0$$

Space Frame Analyses and Slab Membrane Action 133

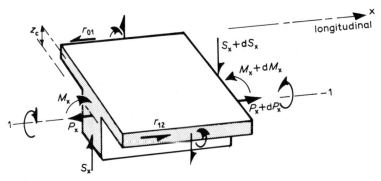

Fig. 7.9 Element of beam of beam-and-slab deck.

and secondly on taking moments about Oy, (7.1)

$$\frac{dM_x}{dx} + \frac{z_c dP_x}{dx} = S_x$$

where z_c is distance of centroid of beam below centroid of slabs.

The axial stress at any level is calculated as for prestress beam from

$$\sigma = \frac{M_x z}{I_x} + \frac{P_x}{A_x}. \qquad (7.2)$$

Fig. 7.10 gives an example of the downstand grillage output for longitudinal and transverse members at a cross-section of a deck. The axial (bending) stresses derived from this output using Equation 7.2 are shown in Fig. 7.11.

The web shear forces and beam torsions in longitudinal members are precisely as output.

The shear flow in the flanges (ignoring inter-beam shear for the moment) is found as outlined for a beam in Section 2.3.2, except that the change in tension along the fibre element of Fig. 2.6 now includes a component due to gross axial force P in the beam;

$$r_f = \tau t = \frac{dM_x}{dx} \frac{A\bar{z}}{I_x} + \frac{dP_x}{dx} \frac{A}{A_x} \qquad (7.3)$$

where $\bar{z}$ is the distance of centroid of element area A of cross-section from centroid of section. A_x and I_x are the area and second moment of area of the beam.

Using Equation 7.1 we obtain

$$r_f = \tau t = \frac{S_x A\bar{z}}{I_x} - (r_{12} - r_{01}) \left[\frac{A}{A_x} - \frac{z_c \bar{z} A}{I_x} \right]. \qquad (7.4)$$

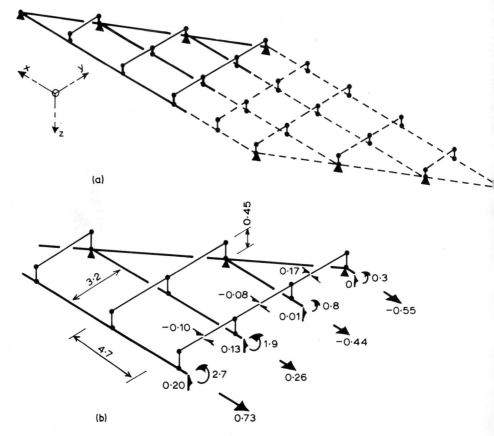

Fig. 7.10 Downstand grillage of deck with section of Fig. 7.11

The inter-beam shear flows r_{12} and r_{01} can be calculated directly from the output by dividing the in-plane shear forces in transverse members in Fig. 7.10 by the width of member each represents. They are shown in Fig. 7.12a plotted as dashed lines. Using Equation 7.4 and the output web shear forces S_x, we can calculate the flange shear flows at the edges of the web. The flange shear flows have triangular distributions decreasing from maximum values at the webs to zero at the edges of the flanges of the T-beams. They are shown in Fig. 7.12a superimposed on the inter-beam shear flow to give the total shear flow picture.

7.3 McHenry lattice space frame

The downstand grillage described above is restricted to beam-and-slab decks for which sideways deflections of the slab are negligible. A less restricted but more complicated space frame model can be made using a McHenry lattice.

Space Frame Analyses and Slab Membrane Action 135

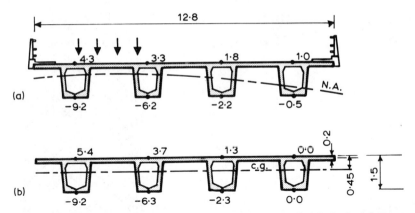

Fig. 7.11 Axial (bending) stresses in spaced beam-and-slab deck. (a) From downstand grillage (b) from plane grillage.

McHenry [1] showed that the in-plane deformation of a plate such as Fig. 7.13a could be investigated using an equivalent lattice such as in (b). Each element of plate [in (c)] is represented by a pin jointed lattice [in (d)] with the equivalent member widths shown. With a fine mesh the lattice can accurately reproduce the behaviour of a plate having Poisson's Ratio $\nu = 1/3$. The error for other values of ν found in common construction materials is generally negligible.

A McHenry lattice can be used to make a space frame model of a beam-and-slab deck as in Fig. 7.14a. The slab joints are not restrained as in downstand grillage. The lattice can also be used for analysis of a cellular deck as in Fig. 7.14b. However, as shown by Hook and Richmond [2], a shear flexible grillage can also

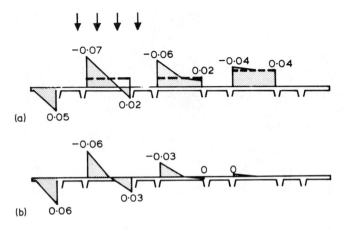

Fig. 7.12 Slab shear flows in spaced beam-and-slab deck. (a) From downstand grillage (b) from plane grillage.

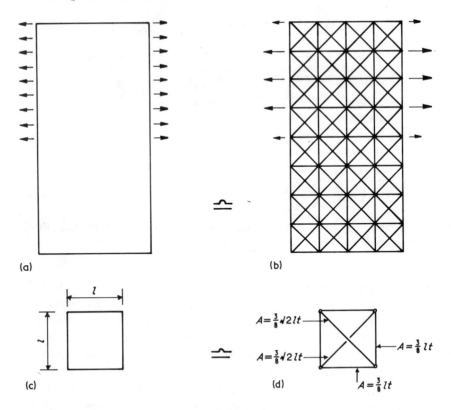

Fig. 7.13 McHenry Lattice.

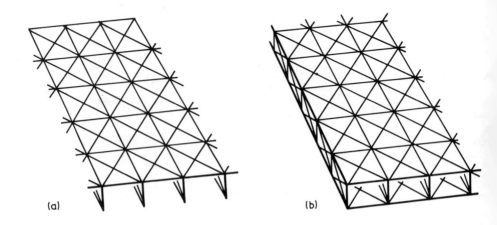

Fig. 7.14 McHenry Lattice space frames.

be used with similar accuracy for a cellular deck. This is because the inter-beam shear in the top slab is balanced by an opposed shear force in the bottom slab (rather than by increasing tension or compression in beam). The opposed in-plane shear stiffnesses of top and bottom slab can be adequately represented by the cell torsion constants described in Chapter 5. However, the McHenry Lattice does provide a better representation of the intersections between diaphragms and webs than is possible with a plane grillage. Furthermore, since the lattice elements can represent plates in any plane, the McHenry Lattice space frame is much more versatile and accurate for the analysis of decks with inclined webs or with numerous or skew diaphragms.

7.4 Cruciform space frame

Beam-and-slab and cellular bridge decks can also be modelled in three dimensions with a space frame which has its members arranged in cruciforms, as illustrated in Fig. 7.15. This method, as versatile as the McHenry lattice method and simpler, was developed by British Rail [3]. The slabs and webs of the prototype are notionally subdivided into rectangular elements (of as nearly the same size as possible), and each of these is represented in the space frame model by a cruciform of beam members as shown in Fig. 7.16. Since there is no interaction in the cruciform between compression and bending in one direction with the compression and bending in the other direction, the model effectively assumes that Poisson's Ratio is zero (except where used in the shear modulus).

A typical member PQ of the cruciform with the orientation of Fig. 7.16 is

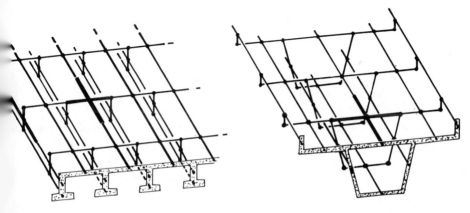

Fig. 7.15 Cruciform space frames.

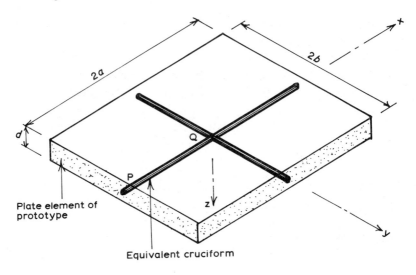

Fig. 7.16 Cruciform element.

given section properties about local member axes:

Compression area $\quad A_{xx} = 2bd$

In-plane shear area $\quad A_{yy} = \dfrac{2bd}{[1 - 0.4(a/b)^2]}$

Out-of-plane shear area $\quad A_{zz} = 2bd$ (7.5)

Torsion constant $\quad C_{xx} = \dfrac{2bd^3}{6}$

Out-of-plane bending inertia $\quad I_{yy} = \dfrac{2bd^3}{12}$

In-plane bending inertia $\quad I_{zz} = \dfrac{d(2b)^3}{12}$

It can be seen that the expression for shear area A_{yy} is a hyperbolic function, and that it tends to infinity for a/b close to 1.58 and is negative for $a/b > 1.58$. Unless A_{yy} can be input as rigid, this ratio of $a/b = 1.58$ should be avoided.

The cruciform model can also be used with members in all three directions to simulate a solid block of material. In this case, both shear areas of each member are calculated as for A_{yy} above.

7.5 Effects of slab membrane action on beam-and-slab deck behaviour

7.5.1 Axial stresses and movement of neutral axis

The axial stress at any point is a combination of stress due to bending moment M_x and tension P_x. Figs. 7.11a and 7.17a show the axial stresses on the cross-sections of decks with beams spaced and contiguous respectively. The neutral axis, where the combined bending and direct stress is zero, moves up in regions of deck subjected to load and downwards elsewhere.

For comparison, Figs. 7.11b and 7.17b show the stresses calculated from bending moments in a plane grillage analysis. It is interesting to note that it is only the slab stresses which are significantly affected by slab membrane action. The soffit stresses are similar from the two analyses. However if the deck also has continuous upstand parapets, the movement of the neutral axis is much more significant at the edges, as discussed in Chapter 8.

7.5.2 Slab shear flows

Fig. 7.12b shows the slab shear flows calculated from the plane grillage corresponding to the downstand grillage of (a). It is evident that for this spaced

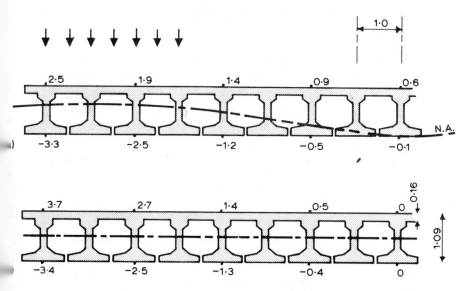

Fig. 7.17 Axial (bending) stresses in contiguous beam-and-slab deck. (a) From downstand grillage (b) from plane grillage.

140 *Bridge Deck Behaviour*

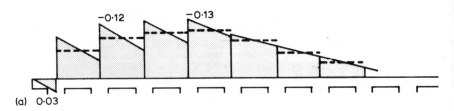

Fig. 7.18   Slab shear flows in contiguous beam-and-slab deck. (a) From downstand grillage   (b) from plane grillage.

beam deck the inter-beam shear flow is small and so the maximum in-plane shear from the plane grillage does not differ significantly from the downstand grillage. However this is not the case for contiguous beam-and-slab decks, as shown in Fig. 7.18a and b for downstand and plane grillage respectively. Because the beams are so close, the slab strips between are very stiff in shear, and as a result the inter-beam shear flow forms a high proportion of the total. In the case of Fig. 7.18, the shear flow in the downstand grillage is three times that calculated from the plane grillage.

An approximate estimate of the high shear flow in the slab of contiguous beam-and-slab decks can be obtained from plane grillage output for the usually critical section at the edge of the loaded area. Slab membrane action, as discussed in Section 7.5.1, only has significant effect on the slab and not on the bottom flanges. As far as the beam at the edge of the loaded area is concerned, the loadfree deck to the side behaves in membrane action like a very wide flange as shown in Fig. 7.17. The effective width of this flange is reduced by shear lag

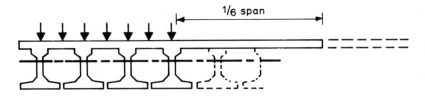

Fig. 7.19   Effective flange section for slab shear flow at edge of load.

to about 1/6 span. By recalculating the section properties of the deck under the load with such wide flanges and using the plane grillage output S_x (which differs little from S_x of downstand grillage), simple beam theory Equation 2.4 can be used to predict values of shear flow in the slab similar to those from the downstand grillage. It can be seen that the wider the beam spacing, the less significant the 1/6 span is compared to the flange associated with each beam, and thus slab membrane action has less significance.

REFERENCES

1. McHenry, D. (1943), 'A lattice analogy for the solution of stress problems,' *J. Inst. Civ. Eng.*, **21**, pp. 59–82.
2. Hook, D. M. A. and Richmond, B. (1970), 'Precast concrete box beams in cellular bridge decks,' *Struct. Eng.*, **48**, pp. 120–128.
3. British Railways Board (1972), 'Computer analysis of plates and slabs using frame and grillage programs and required data preparation,' Technical Note No. 26, Civil Engineering Department. Internal report.

8
Shear lag and edge stiffening

8.1 Shear lag

The thin slabs of cellular and beam-and-slab decks can be thought of as flanges of I- or T-beams as shown in Fig. 8.1. When such I- or T-beams are flexed, the compression/tension force in each flange near midspan is injected into the flange by longitudinal edge shear forces, shown in Fig. 8.2. (This Fig. also shows the coexistent transverse in-plane forces which prevent the flanges on each side of a web flexing away from each other.) Under the action of the axial compression and eccentric edge shear flows, the flange distorts (as in Fig. 8.3) and does not compress as assumed in simple beam theory with plane sections remaining plane. The amount of distortion depends on both the shape of the flange in plane and on the distribution of shear flow along its edge. As is evident in Fig. 8.3a, a narrow flange distorts little and its behaviour approximates to that assumed in simple beam theory. In contrast, the wide flanges of (c) and (d) distort seriously

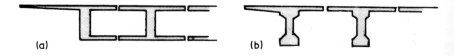

Fig. 8.1 Deck sections divided into I- or T-beams.

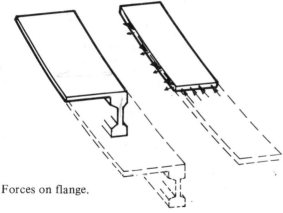

Fig. 8.2 Forces on flange.

because the compression induced by the edge shears does not flow very far from the loaded edge, and much of each wide flange is ineffective. The decrease in flange compression away from the loaded edge due to shear distortion is called 'shear lag'.

8.2 Effective width of flanges

To enable simple beam theory to be used for analysis of beams with wide flanges, the flanges are attributed 'effective flange widths.' The effective width of a flange is the width of a hypothetical flange that compresses uniformly across its width by the same amount as the loaded edge of the real flange under the same edge shear forces. Alternatively, the effective width can be thought of as the width of theoretical flange which carries a compression force with uniform stress of magnitude equal to the peak stress at the edge of the prototype wide flange when carrying the same total compression force. Fig. 8.3 shows the

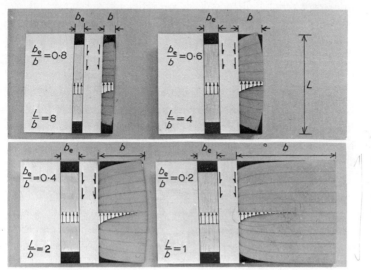

Fig. 8.3 Shear lag distortion of flanges of various widths.

144 Bridge Deck Behaviour

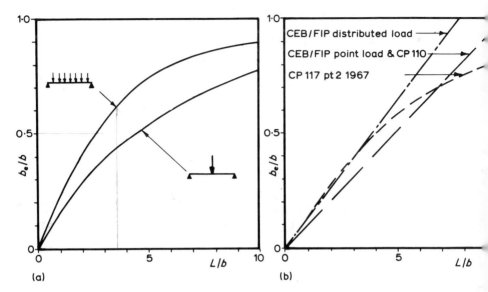

Fig. 8.4 Effective widths of flanges of various shapes. (a) Theoretical expressions for point and distributed loads (b) recommendations of Codes of Practice.

effective widths of the particular shapes of flange and the distributions of midspan compression stress in each prototype and its theoretical effective equivalent.

General relationships between effective flange width and length of flange between points of zero compression (i.e. contraflexure) are shown in Fig. 8.4. (a) indicates theoretical relationships between the effectiveness b_e/b and shape L/b for flanges of beams supporting either a uniformly distributed load or midspan point load. (b) shows the relationships recommended by various Codes of Practice. The difference in effective width of flange resulting from different distributions of load can be explained with Fig. 8.5. Under action of a point load in (a), the flange edge shear flow is large right up to the load, and the compression induced by shear flows near midspan cannot spread far across the flange. In contrast, the shear flows for the distributed load in (b) are predominantly applied at the ends of the flange, and the compression they induce has most of the length of the flange to spread out. A rigorous analysis of the effects of load distribution and flange shape is extremely complicated. While the relationships of the Codes in Fig. 8.4b are generally sufficiently accurate for analysis of concrete bridges, reference [1] contains much more complex requirements for steel box girder bridges. The calculation of effective flange widths of thin steel plates is more complicated than for concrete sections

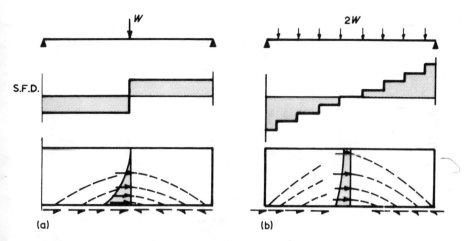

Fig. 8.5 Different stress distributions in flange for (a) concentrated load and (b) distributed load.

because the effectiveness of a thin flange is reduced by plate buckling as well as by shear lag. Naturally, if a detailed three-dimensional analysis is carried out using a lattice, folded plate or finite element model which considers slab membrane behaviour, the effects of shear lag are automatically included and it is not necessary to determine what the effective width of a flange is.

Continuous bridge decks have significant variations in effective flange width along their length. This is partly due to the different distances between points of contraflexure along spans and over supports, and partly due to the very different dominating loads distributed over spans and concentrated at support reactions. Fig. 8.6 demonstrates the variations of edge shear flow along the length of a deck, and of spread of compression or tension across the flange.

The effects of shear lag are incorporated in grillage analysis by reducing the flanges associated with each longitudinal beam in accordance with Fig. 8.4. Fig. 8.7 shows part cross-sections of two decks and the effective sections used for calculation of grillage member properties. These sections are also used for calculating from the grillage output the values of the peak bending stresses at webs and the values of the shear flows at the roots of the flanges.

The true distribution of longitudinal bending stress in a wide flange decreases towards the outside edge. Its shape can be estimated from the peak value, derived from grillage, and the effective width of the flange. Fig. 8.8 shows the effective section for grillage of an elemental I-beam of a cellular deck. Applying simple beam theory to the reduced section, the stress is uniform across each effective flange (as shown dotted) and equal to the actual peak stress at the web edge. The true stress distribution decays away from the peak value and 'flattens

146 *Bridge Deck Behaviour*

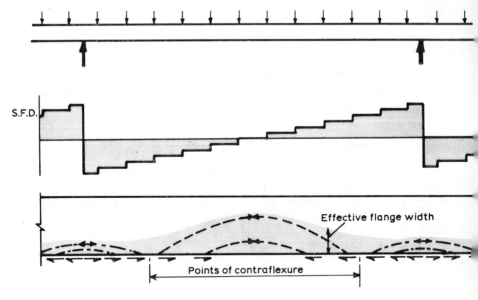

Fig. 8.6   Varying effective width of flange of continuous deck.

out' towards the real edge of the flange. The total axial force in each flange, i.e. area under stress curve, is the same for uniform narrow effective flange and wider actual flange. Hence the true stress distribution can be sketched to pass through the peak value and enclose the same area as the uniform stress on the reduced flange.

8.3 Edge stiffening of slab decks

A slab deck is better able to carry a load near an edge if the edge is stiffened with a beam. Fig. 8.9a shows a slab deck with edge stiffening beams which have their centroids on the mid plane of the slab. The bending inertias of such beams are calculated about the mid plane of the slab and the beam sections are fully effective. Improved edge stiffening is achieved if the beams do not have their

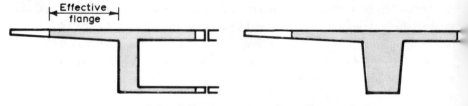

Fig. 8.7   Effective sections for grillage analysis.

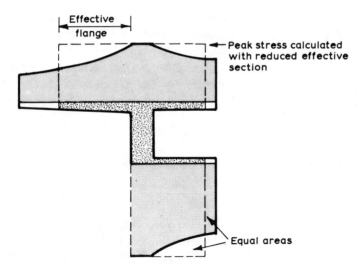

Fig. 8.8 Determination of bending stress distribution from peak value and effective flange width.

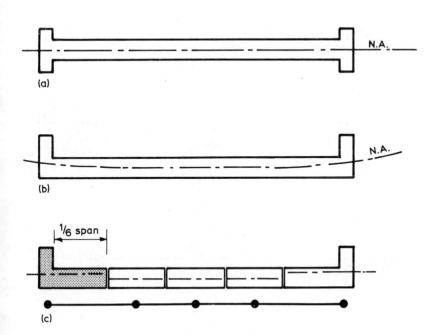

Fig. 8.9 Edge stiffening of slab. (a) Edge beam centroids on midplane of slab (b) edge beam centroids above midplane of slab (c) sections for grillage.

148   Bridge Deck Behaviour

centroids on the mid plane of the slab as in (b) because the beams then act as L-beams with the slab deck acting to some extent as a flange. Under bending action, the neutral axis remains near the mid plane of the slab in central regions and rises towards the edges. The width of the slab that acts as flange to the edge beam is restricted by the action of shear lag. The effective width can be determined as described in Section 8.2 and is approximately 1/6 of the span as shown in Fig. 8.9c.

8.4 Upstand parapets to beam-and-slab decks

The load distribution characteristics of a beam-and-slab deck can be greatly improved by making the parapet part of the structure. Fig. 8.10a and b show the bending stresses computed from a folded plate analysis (see Chapter 12) of decks with and without upstand parapets supporting loads near the edge beams. It is evident that while the top of the upstand attracts a high compressive stress, the accompanying stresses in the edge main beam are much smaller than those in the deck without structural parapet. The parapet effectively acts with the edge main beam, as shown in Fig. 8.11.

The predicted slab stresses in Figs. 8.10 and 8.11 differ because in the grillage

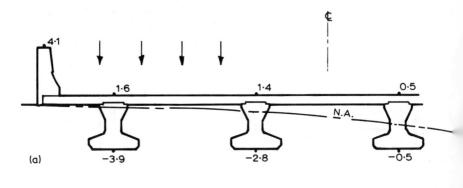

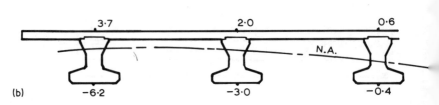

Fig. 8.10 Bending stresses in beam-and-slab deck (a) with and (b) without structural parapet. (From folded plate analysis.)

Shear Lag and Edge Stiffening   149

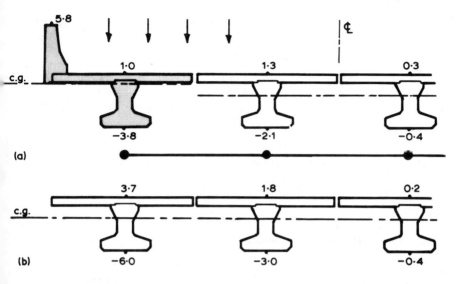

Fig. 8.11   Bending stresses predicted by grillage for deck of Fig. 8.10.

analysis the slab to the right of the loaded beam is not subjected to additional compression by inter-beam shear as described in Chapter 7. Furthermore, but in contrast, by assuming in the grillage that the parapet is part of the edge beam, shear lag deformation of the thin cantilever slab has been ignored and the parapet appears more effective, thus attracting higher stresses. Such comparison between a plane grillage and a three-dimensional structural analysis is highly dependent on the closeness of the beams. If the beams are close together, several of them will act compositely with the upstand parapet. Consequently, if the parapet is assumed part of the structural edge beam in a plane grillage analysis, it is advisable to check the effectiveness of the section by comparing the level of its centroid with the level of the neutral axis derived from a three-dimensional analysis (downstand grillage, folded plate or finite element).

The disadvantages of making the parapet structural often outweigh the benefits. Firstly, the ends of the parapet must be properly supported on diaphragm beams. Furthermore, construction sequence can be much more critical and the effects of differential shrinkage are more severe. The structural integrity of the parapet and joints is essential, and the buckling stability of the parapet top in compression must be checked. The cantilever slab is subjected to longitudinal shear flow forces two or three times those in the free cantilever. The parapet must be so strong that it could not be broken by the impact of a vehicle. Finally, the analysis must be carried out with much more care. A conclusion is that if upstand parapets are required for traffic reasons, but not structurally,

150 *Bridge Deck Behaviour*

they should be made discontinuous with frequent expansion joints (at the same time they must retain sufficient transverse strength to prevent vehicles passing through).

Upstand parapets can also be used structurally on cellular decks, but the disadvantages are generally even more pronounced. The load distribution characteristics of a cellular deck are usually so good that the whole cross-section, or a wide part of it, is effective in supporting a load near an edge. The addition of a tall upstand has little effect on stiffness while it attracts high compressive stresses to the top of the upstand and high shear stresses to the cantilever slab. However, a small beam on the outside edge of a wide cantilever can be useful in providing local stiffening to the edge.

8.5 Service bays in beam-and-slab decks

Services are often carried on bridges in service bays placed under the footways or verges. Some are open at the top along their full length, (but spanned by paving

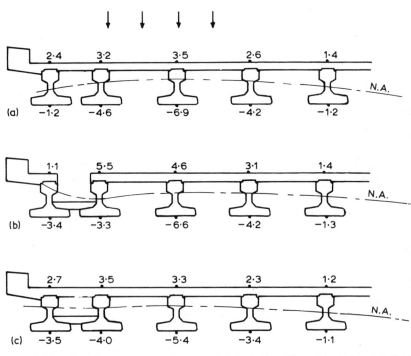

Fig. 8.12 Bending stresses in beam-and-slab deck with various edge details. (a) Without service bay (b) open service bay (c) box service bay. (From folded plate analysis.)

Shear Lag and Edge Stiffening   151

slabs) as shown in Fig. 8.12b, while others are effectively box beams as in Fig. 8.12c. The structural actions of these different structures when loads are placed near them are very different. Fig. 8.12 shows the bending stresses computed from folded plate analysis of the three different structural forms. It is evident that the best distribution (i.e. with lowest maximum stresses) is obtained when the edge beams behave as a box. However, the shear force in the web next to the edge is then large and often critical for design, since in addition to attracting a high bending shear it is subjected to high torsional shear flows. It should be noted that the box will only be torsionally stiff if the bottom slab is effectively connected to the web on each side for transfer of longitudinal shear flow. In practice, it is very difficult to construct a structurally effective *in situ* concrete slab to form the bottom of the trough or box between precast beams, as the joints are unlikely to have sufficient stiffness in either bending or longitudinal shear.

The deck with open trough service bay has the worst distribution because the

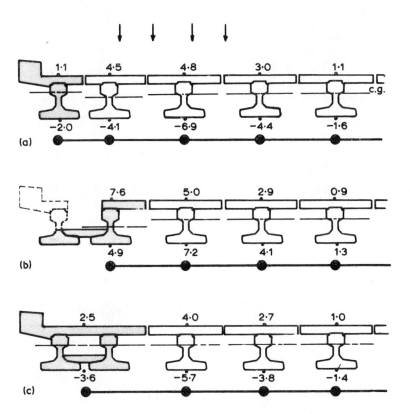

Fig. 8.13   Bending stresses predicted by grillage models of decks of Fig. 8.12.

trough behaves as a U-beam subjected to eccentric load and the outside web is largely ineffective. This inefficiency can be largely avoided while retaining the benefit of easy access by giving the service bay the box cross-section near midspan and the open trough section towards the ends. The high transverse bending stiffness of the box at midspan provides effective load distribution where it is needed, while the removal of the top slab near the supports avoids high torsions in the edge box which overstress the web next to the edge in shear. Finally, it should be noted that if the stiffness and strength of the outside beam is ignored in the design, the slabs connecting this beam to the deck should be designed to articulate. If they are made continuous but only given nominal strength, they will break when the deck flexes.

Fig. 8.13 shows, for comparison, the bending stresses computed from various grillage models of the decks in Fig. 8.12. The different sections assumed for the edge beam are shown shaded.

REFERENCES

1. Department of the Environment (1973), 'Inquiry into the basis of design and method of erection of steel box girder bridges,' Report of the Committee, London, Her Majesty's Stationery Office.

9
Skew, tapered and curved decks

9.1 Skew decks

9.1.1 Characteristics of skew decks

The majority of bridge decks built today have some form of skew, taper or curve. Because of the increasing restriction on available space for traffic schemes and also due to the increasing speed of the traffic, the alignment of a transport system can seldom be adjusted for the purpose of reducing the skew or complexity of the bridges. Fortunately this increasing demand for high skew bridges has been accompanied by the development of computer aided methods of analysis, and it is now generally possible to design a structure at any angle of skew.

In addition to introducing problems in the design of details of a deck, skew has a considerable effect on the deck's behaviour and critical design stresses. The special characteristics of skew of a slab deck are summarized in Fig. 9.1. They are:

(1) Variation in direction of maximum bending moment across width, from near parallel to span at edge, to near orthogonal to abutment in central regions
(2) hogging moments near obtuse corner
(3) considerable torsion of deck
(4) high reactions and shear forces near obtuse corner
(5) low reactions and possibly uplift in acute corner.

154 *Bridge Deck Behaviour*

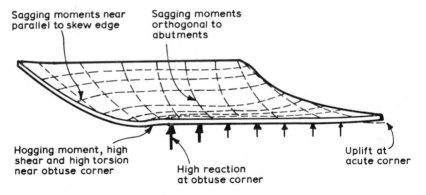

Fig. 9.1   Characteristics of skew slab deck.

The size of these effects depends on the angle of skew, the ratio of width to span, and particularly on the type of construction of the deck and the supports. Fig. 9.2 shows how the shape and edge details can influence the direction of maximum moments. While in (a) and (b) the decks span onto the abutments, in (c) the stiff edge beam acts as a line support for the slab which effectively spans right to the abutment across the full width. In (d) the skew is so high that the decks is cantilevered off the abutments at the acute corners.

The deleterious effects of skew can be reduced by supporting the deck on

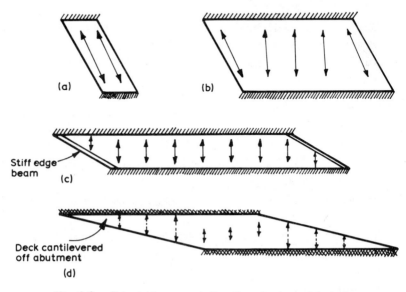

Fig. 9.2   Principal moment directions in skew slab decks.

soft bearings. The high reaction on the bearing at the obtuse corner is shed to neighbouring bearings. In addition to reducing the magnitude of the maximum reaction, this also reduces the shear stresses due to shear and torsion in the slab and it reduces the hogging moment at the obtuse corner. Uplift at the acute corner can also be eliminated. However, this redistribution of forces along the abutment is accompanied by an increase in sagging moment in the span.

The above characteristics are particularly significant in solid and cellular slab decks because their high torsional stiffness tries to resist the twisting of the deck. In contrast, skew is less significant in beam-and-slab decks, particularly with spaced beams. Fig. 9.3a, b and c shows a plan, elevation and right section of a spaced beam-and-slab deck subjected to uniform load. At the abutments there is a large difference in longitudinal slope at adjacent points on neighbouring beams [evident in (b)] and also a relative vertical displacement [evident in (c).] This distortion of the deck can occur without generating large reactive forces if the torsional stiffnesses of the slab and beams are low. Under the action of a local concentrated load, distribution still takes place by transverse bending of the slab, but the beams behave much as in a right deck spanning longitudinally. However, the increase in beam shear force and reaction at the obtuse corner is still significant and should be considered. Uplift at the acute corner is unlikely. It should be noted that if the beams have box section with high torsional stiffness, they will attract high torques. It may well be found that the torsion shear in the webs is then excessive, and torsionally flexible I-beams may be more appropriate.

The effects of skew are generally considered negligible for simply supported decks with skew angle less than 20°. However, the effects are significant at lower skew angles in continuous decks, particularly in the region of intermediate supports. Fig. 9.4a and b shows grillage bending moment diagrams for the edge

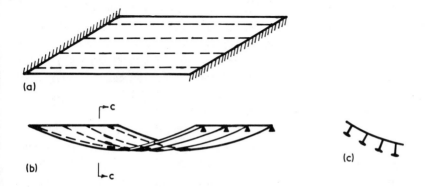

Fig. 9.3 Skew beam-and-slab deck. (a) Plan   (b) elevation   (c) section.

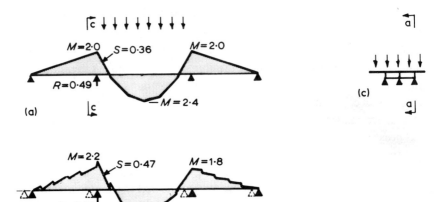

Fig. 9.4 Grillage moment diagram for edge web of (a) right and (b) 20° skew three span cellular decks.

web of a three span cellular deck with right supports in (a) and 20° skew in (b). Both are loaded over the centre span. There is little difference in midspan moments. However at the support skewed towards the loaded span the moment, shear force (slope of saw tooth moment diagram) and reaction are all greatly increased by the skew.

9.1.2 Grillage meshes for skew decks

Design moments in simply supported skew isotropic slab decks can be obtained from the influence surfaces of Rusch and Hergenroder [1] or of Balas and Hanuska [2]. However, these charts have the disadvantages that they are difficult to use, do not give the user a complete picture of the force system in the deck under a particular load case, and cannot be used for orthotropic, cellular or beam-and-slab decks because of their very different distortional and torsional characteristics. In general, a grillage analysis is much more convenient for all types of deck. Even during the preliminary design stage when it is not clear what span-to-depth ratio is appropriate to the skew and method of construction, a preliminary quick crude grillage is preferable to interpretation and conversion of the charts.

A skew deck can be analysed with a grillage having either a skew mesh as in Fig. 9.5a or orthogonal mesh as in (b) or (c). While the skew mesh is convenient for low skew angles, it is not appropriate for angles of skew greater than 20° because it has no members close to the direction of dominating structural action. If a deck does have a higher skew angle with reinforcement parallel to these skew

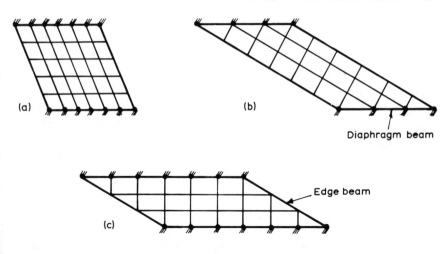

Fig. 9.5 Grillages for skew decks. (a) Skew mesh (b) mesh orthogonal to span (c) mesh orthogonal to support.

directions, the quantity of reinforcement is likely to be excessive and uneconomic. In general, the grillage members should be placed parallel to the designed lines of strength, which are usually orthogonal. Clark [3] shows that for narrow slab decks the reinforcement should be placed in directions of mesh of (b), while for decks with abutment length greater than span, the directions of (c) are generally more appropriate. However, for cellular decks and beam-and-slab decks, the longitudinal grillage members should be parallel to the webs or beams which are usually as in (b).

While the general comments of Chapters 3, 4 and 5 apply to skew decks, a problem arises in relation to member properties near the diaphragm beams in Fig. 9.5b and edge beams in (c). It might appear that if members are given uniform properties, the parts of the deck in the triangular regions are represented more than once. If the orthogonal members are assumed to represent the typical deck construction, they should have the typical section properties. No reduction in the grillage member stiffness is necessary since the strip of prototype it represents is not tapered and does not terminate at a point. The edge trimming grillage member should be given the stiffness appropriate to the additional stiffness coincident with its line in the prototype. If the prototype is a slab with no special end or edge stiffening, the skew grillage member is given nominal stiffness. However, if there is a stiffening diaphragm or edge beam built into or onto the deck, the equivalent grillage member should be given the stiffness appropriate to the dimensions and strength of the actual stiffening member with the part of slab that participates as its flange.

Even though the lines of design strength and grillage member may be chosen to be near parallel to the principal moment directions, the grillage might still predict high torques in places. The maximum moments and stresses in slab decks are then calculated from the orthogonal moments and torques by using the equations of Section 3.3.3. The reinforcement of concrete decks must be designed to resist the combinations of moment and torque, and the equations of Wood [4] and Armer [5] are relevant. The grillage output of beam-and-slab and cellular decks is interpreted in the ways described in Sections 4.7 and 5.7 respectively.

9.2 Tapered decks

Bridge decks seldom have a very pronounced taper, and a tapered grillage as in Fig. 9.6 can be used without special thought. The only problem is that grillage member properties must be incrementally increased along the strings of members, thus making data preparation and output interpretation cumbersome. Often a small taper can be ignored in the analysis as it has little effect on the deck's behaviour. However, if a deck also is at a high skew angle, the edges will have very different spans and both taper and skew must be reproduced.

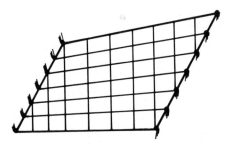

Fig. 9.6 Tapered grillage.

A fan type structure, as Fig. 9.7, can be analysed with a grillage. The subtended angle between neighbouring radial members can be as large as 15° without introducing significant error. The curvilinear squares of the grillage mesh should be near 'square', and the radial members given the stiffness equivalent to the section midway along their length. It might appear that the mesh should get very fine close to the centre. But when thought is given to how the lines of strength in the prototype, in the form of reinforcement or beams cannot taper to a point but must be curtailed, it will be evident that the grillage mesh should also be curtailed.

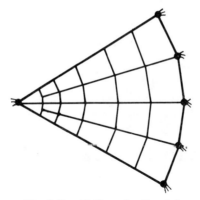

Fig. 9.7 Grillage for fan slab.

9.3 Curved decks

9.3.1 Curved beams

When a vertical load is placed on a curved beam there is an interaction of moment and torsion along the length of the beam. If the angle of curvature between supports is less than 20°, the effect of the torsion on the bending moment is small and the deck can be considered straight with spans equal to the arc lengths. However, special consideration needs to be given to bearing reactions. In general, the simplest method of analysis, particularly for several load cases, is to use a grillage and represent a curved beam by a 'curved' string of straight members as described below. Nonetheless convenient approximate hand methods are described by Witecki [6] and Toppler *et al.* [7]. In multi-beam decks, interaction between moment and torsion is significant at lower angles of curvature due to redistribution between beams. Before discussing grillage analysis of curved decks, the equilibrium and stiffness equations for a curved beam will be presented.

Fig. 9.8 shows an element of beam curved in plan. It has length ds, radius of plan curvature r, and subtends an angle $d\alpha$. It is subjected to an element of vertical load dW at eccentricity y; this force is resisted by the beam moment M, shear force S and torsion T. Equilibrium of the element requires:

$$dS = -dW = -W(s)\,ds$$

$$\frac{dM}{ds} - \frac{T}{r} = S \qquad (9.1)$$

$$\frac{M}{r} + \frac{dT}{ds} = y\frac{dW}{ds}.$$

160   *Bridge Deck Behaviour*

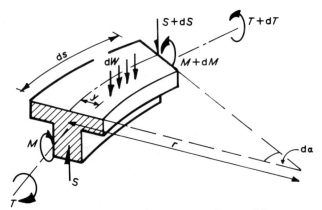

Fig. 9.8   Forces on element of curved beam.

The flexural and torsional stiffness of the element are the same as those for the straight beam in Equations 2.5 and 2.14 except that moment and torsion are related to deflections by

$$M = -EI \left(\frac{d^2 w}{ds^2} - \frac{\phi}{r} \right) \qquad T = -GC \left(\frac{d\phi}{ds} + \frac{1}{r} \frac{dw}{ds} \right). \tag{9.2}$$

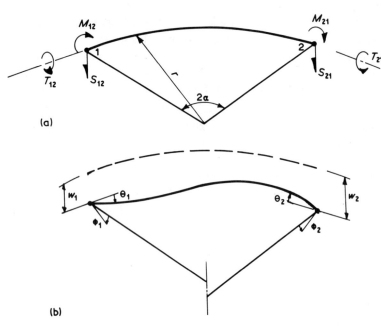

Fig. 9.9   (a) End forces and (b) end displacements on curved beam.

Skew, Tapered and Curved Decks

If the flexural and torsional rigidities EI and GC are uniform along the beam, the end forces and end displacements of Fig. 9.9 are related by the complex slope-deflection equations given in reference [8].

9.3.2 Grillage analysis of curved decks

A curved bridge deck can be represented for the purpose of analysis by a grillage composed either of curved members as in Fig. 9.10a, or of straight members as in (b). While some computer programs do have the facility to represent curved members, the improvement in accuracy over the straight member grillage is not significant enough to warrant its general use. Most programs do not have the facility and straight members must be used in any case.

It will be found that if the general recommendations of Chapters 3, 4 and 5 are used to determine member spacing, the maximum change in direction at a joint will seldom need to be more than 5°. This is much smaller than the angle at which the behaviour of a true curved beam differs significantly from the 'curve' of straight members. In the prototype, moment and torsion interact continuously and smoothly along a curved member. In the straight member grillage such interaction only occurs at joints, so that each type of force is discontinuous. However, the values of forces midway along members are representative of those in the prototype. A smooth distribution can be plotted through the values at these points, and from it the values elsewhere interpolated. The interpretation of the grillage output is precisely the same as that described in Chapters 3, 4 and 5 for the relevent type of construction.

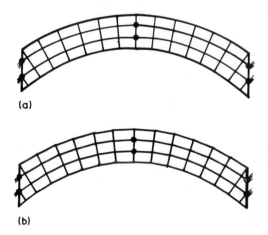

Fig. 9.10 Grillages of curved decks. (a) Curved members (b) straight members.

REFERENCES

1. Rusch, H. and Hergenroder, A. (1961), *Influence Surfaces for Moments in Skew Slabs*, Munich, Technological University, 1961. Translated from German by C. R. Amerongen. London, Cement and Concrete Association.
2. Balas, J. and Hanuska, A. (1964), *Influence Surfaces of Skew Plates*, Vydaratelstvo Slovenskej Akademie Vied, Bratislava.
3. Clark, L. A. (1970), 'The provision of reinforcement in simply supported skew bridge slabs in accordance with elastic moment fields,' Cement and Concrete Association, Technical Report.
4. Wood, R. H. (1968), 'The reinforcement of slabs in accordance with a predetermined field of moments,' *Concrete*, 2, 69–76.
5. Armer, G. S. T. (1968), 'Discussion of reference 4,' *Concrete*, 2, 319–320.
6. Witecki, A. A. (1969), 'Simplified method for the analysis of torsional moment as an effect of a horizontally curved multispan continuous deck,' First International Symposium on Concrete Bridge Design. A.C.I. Publication SP-23, pp. 193–204.
7. Toppler, J. F., Chaudhuri, B. K., van den Berg, J. and Harris, F. R. (1968), 'Horizontally curved members: an approximate method of design,' *Concrete*, 2, pp. 418–425.
8. Blaszkowiak S. and Kaczkowski, Z. (1966), *Iterative Methods in Structural Analysis* (translated by Kacner, A. and Olesiak, Z.), Pergamon Press, Oxford.

10
Charts for preliminary design

10.1 Introduction

Prior to the general use of the computer aided grillage, charts provided the most convenient method of load distribution which was quick enough for general design use. Finite difference relaxation methods by hand computation were available but were too cumbersome for all but large and complex structures. Numerous charts have now been published for hand calculation of critical design moments, etc. in isotropic and orthotropic slabs and other types of deck. This chapter reviews some of the previously published charts and then demonstrates the use of three charts for preliminary design of slab, beam-and-slab and cellular decks.

All charts have limitations. In general, a chart which enables direct calculation of design moment is restricted to an individual type of deck construction subjected to a particular load case, which may not be the most critical. In contrast, a more versatile set of charts which enable calculation for a variety of deck types under a variety of load cases is more difficult to use and requires a considerable amount of interpolation and hand calculation to determine the critical conditions. Often the complexity of instructions discourages the user from taking the trouble.

Many of the charts available enable accurate calculations to be made for longitudinal bending moment in simply supported bridge decks. However, in general, they are not able to represent in detail the wide variety of cross-section construction, and predictions of transverse moments are often unsatisfactory for all but simple slab decks. In addition, few charts enable accurate calculation of shear force at supports, which usually has a different distribution from that of longitudinal bending moment.

The three charts demonstrated in Sections 10.3 to 10.6 have been designed to permit rapid investigation of the load distribution characteristics of a wide variety of types of deck construction, and to enable rapid calculation of the maximum design moments. To make the charts versatile and simple, no explanation is given of differences in secondary characteristics of the various types of deck. Such behaviour can be highly dependent on factors such as skew, continuity and cross-section type, which can all be represented much more satisfactorily in a grillage or other computer aided analysis. These charts are solely intended to help the designer make the initial choice of type of construction and deck dimensions.

10.2 Some published load distribution charts

10.2.1 Isotropic slabs

One of the most useful sets of design charts is the book of influence surfaces for isotropic slabs with various shapes and support conditions produced by Pucher [1]. In addition to being useful for the determination of critical design moments in simply supported right slab decks, these charts also provide one of the simplest methods of determining moments under concentrated loads on secondary slabs of beam-and-slab and cellular decks. For both applications, the presentation of the influence surfaces makes it relatively easy to isolate the critical load position and then calculate design moments.

Skew simply supported isotropic slabs can be analysed by means of the charts of Rusch and Hergenroder [2] or of Balas and Hanuska [3]. While a great deal of valuable information can be derived from these charts for particular shapes of deck, interpolation between charts of different aspect ratios and different skews can be extremely cumbersome and confusing. In general, even for preliminary design, it is quicker and more reliable to carry out a quick grillage analysis, as described in Chapter 9, with estimated properties.

10.2.2 Orthotropic slabs

The most widely used charts for load distribution in orthotropic slab decks are those of Morice and Little [4]. The theoretical basis of these charts is described by Rowe [5] who describes and demonstrates their application in detail. There are two series of charts which give the load distribution of slabs having no torsional stiffness and for slabs having the full torsional stiffness of isotropic slabs. For most bridge decks, interpolation between the sets of charts is necessary and the references demonstrate a simple though somewhat lengthy tabulated procedure.

The above charts, which are based on harmonic analysis demonstrated in Chapter 12, give the distribution of deflections due to the first harmonic of load. Fortunately, the distribution of longitudinal moments due to most design loads approximates closely to that of the first harmonic deflections. However, transverse moments are highly dependent on the local distribution (i.e. higher harmonics) of concentrated loads, and it is necessary to superpose several harmonic components of the transverse moments. The references describe a second more complicated tabular procedure for this analysis.

To avoid the necessity of superposing a number of harmonic components while determining transverse moments, Cusens and Pama [6, 7] developed a set of charts similar to those of references [4 and 5], but with the first nine harmonics of a midspan point load already superposed. While this reduces the quantity of computation for transverse moments it means that the charts specifically apply to point loads near midspan, so precluding superposition of moments etc. for loads in different positions away from midspan. This probably does not matter for the purpose of preliminary design and these charts, as presented in reference [8], have been found very convenient by some designers. Furthermore, they have the additional advantage of covering a wider range of

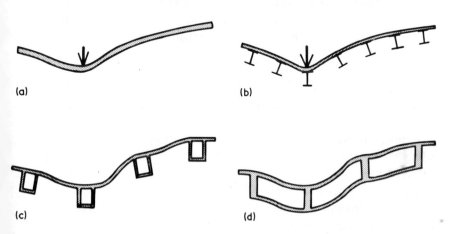

Fig. 10.1 Cross-sections of various decks. (a and b) Distorting in smooth curve (c and d) distorting in series of steps.

torsional stiffnesses which enables them to be used for shear key decks as described in Chapter 6. Shear key decks can also be analysed with the charts of reference [9] which were derived specifically for them.

The charts of references [4–8] all relate to orthotropic slab decks. They can also be used accurately for beam-and-slab and cellular decks if the decks' concentrated stiffnesses can be notionally 'spread out' into a continuum without changing the decks' characteristics. In other words, the charts assume that the deck cross-section deflects in a smooth curve as shown in Fig. 10.1a. If the deck cross-section deflects as in Fig. 10.1c with 'steps' at the concentrations of stiffness, the continuous charts cannot simulate such distortion of the cross-section. This limitation does not apply to the charts demonstrated in the following sections since they are based on the assumption that stiffnesses are concentrated at points across the cross-section so that the deck can distort in 'steps' if appropriate.

10.3 Influence lines for slab, beam and slab and cellular decks

10.3.1 Development of charts

Figs. 10.2–10.4 contain charts for the determination of influence lines for various points across the cross-section of simply supported right decks of slab, beam-and-slab and cellular construction. The charts were developed from repeated application of the approximate folded plate method outlined in Appendix B.

The first step in the analysis of any deck with these charts is to notionally subdivide the deck into a number of parallel 'beams' as shown in Fig. 10.5 and as is necessary for grillage analysis. The physical characteristics of the deck can then be summarized by three non-dimensional parameters which relate the various stiffnesses of the structure. The parameters are:

f the flexural stiffness ratio. This relates the transverse flexural stiffness of the slab or slabs between 'beams' to the longitudinal flexural stiffness of the 'beams'.

r the rotational stiffness ratio. For slab and beam-and-slab decks this relates the torsional stiffness of the slab and beams to the transverse flexural stiffness of the slab. For cellular decks, r relates the flexural stiffness of the webs to the transverse flexural stiffness of the slabs.

c the cellular stiffness ratio. This only applies to cellular decks and relates the cellular torsion stiffness of the deck to the longitudinal flexural stiffness of the 'beams'.

Because the basic equations in Appendix B for slab, beam-and-slab and

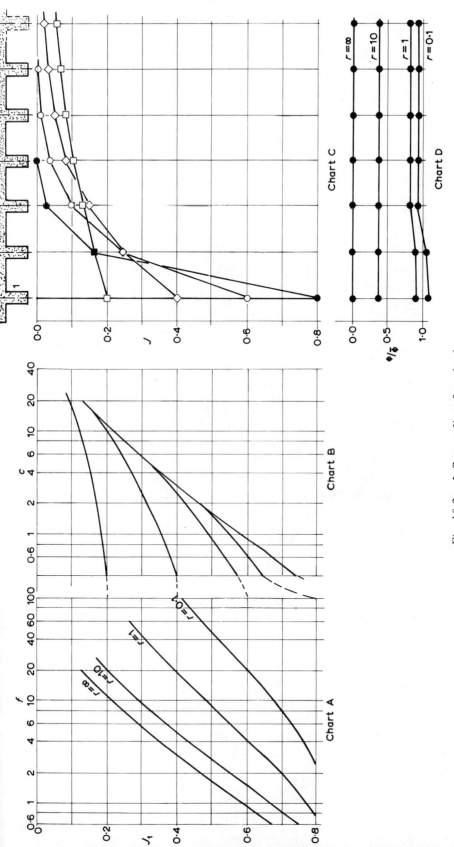

Fig. 10.2 Influence lines for edge beam.

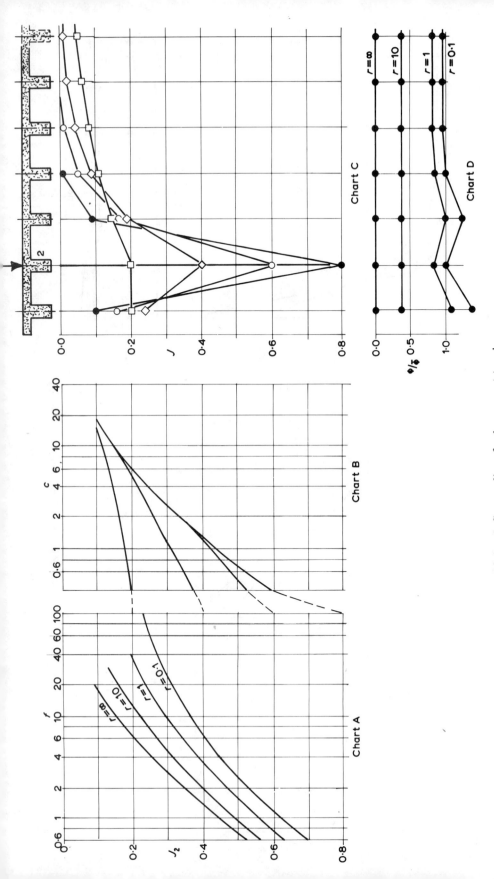

Fig. 10.3 Influence lines for beam next to edge.

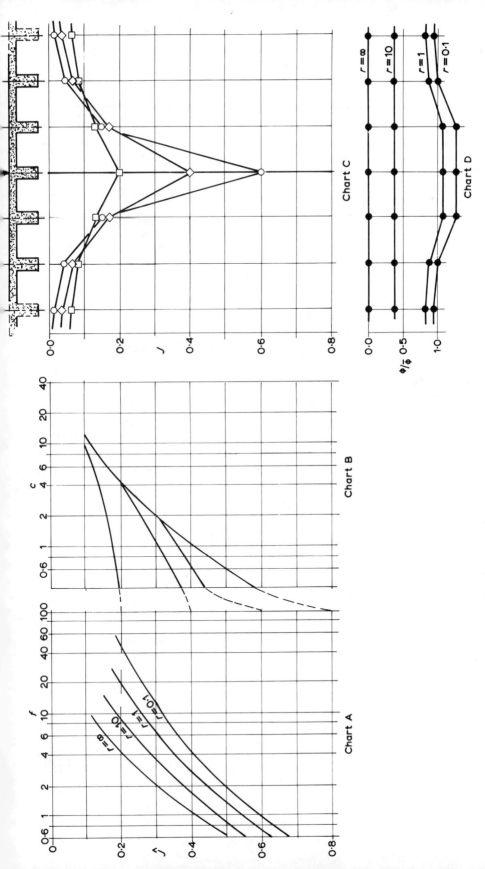

Fig. 10.4 Influence lines for beam far from edge.

170   *Bridge Deck Behaviour*

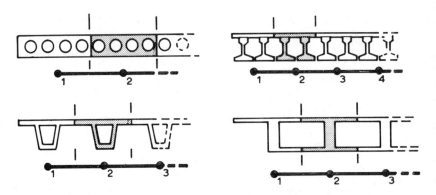

Fig. 10.5   Deck cross-section divided into 'beams'.

cellular decks are similar, it has been possible to derive a single set of charts for all three types of deck. Since the dominating terms in f and r are different for different types of construction, their algebraic definitions are given separately in the demonstrations of the charts in Section 10.4–10.6.

The assumptions of the approximate folded plate theory are given in Appendix B. Some of these are listed below, together with additional assumptions necessary to simplify application of the charts.

(1) The deck is right and simply supported. The charts can also be used for continuous decks for the lengths between points of contraflexure on each span.

(2) The deck is prismatic; i.e. it has the same section from end to end. (If the deck section does vary, the properties near midspan are the most appropriate for approximate use of the charts.)

(3) Slabs and beams are prevented from rotating about a longitudinal axis at their ends by rigid diaphragms.

(4) There are no midspan diaphragms, or the stiffness of such diaphragms is assumed to be 'spread out' along the deck.

(5) The deck has the same transverse bending stiffness at every point across the section. Hence for the beam-and-slab decks of Fig. 10.1 it is assumed that the slab spans transversely between centre lines of 'beams', at which the bending and torsional stiffnesses are assumed concentrated. In the same way, for cellular decks it is assumed that both top and bottom slabs are continuous across the deck and only connected to webs at their midplanes. (If the charts are to be used to analyse a spaced beam-and-slab deck constructed of wide box beams which distort, a supplementary plane frame analysis of the cell distortion is necessary to determine the equivalent uniform slab.)

(6) The deck cross-section can be divided into a number of identical equidistant 'beams'. Edge stiffening can only be represented if it can be thought of as an additional width of the uniform or repetitive section.

Charts for Preliminary Design   171

(7) The centroid of the deck is everywhere at the same level, so that, as in plane grillage analysis, slab membrane action of beam-and-slab decks is ignored.

(8) In cellular decks the ratio of the individual flexural stiffnesses of the top and bottom slabs is within the range 0.3–3. (If the slabs differ by more than this, a check should be made using a plane frame analysis of the section to compare the distortional stiffness of the prototype cell with that of a cell of the same geometry but having identical top and bottom slabs whose combined stiffnesses equal those of the prototype). It is also assumed that the cross-section distorts as in Fig. 10.7b without horizontal sideways deflection of the slabs.

(9) The values of the influence lines at 'beams' to the side of the 'beam' considered decay in geometric progression. In other words, the values of the points on each influence line in Charts C of Figs. 10.2–10.4 decrease in geometric progression away from the peak values. This assumption is discussed later.

Charts A and B in Figs. 10.2–10.4 enable the user to derive the peak value of the influence line for a particular 'beam' from the non-dimensional parameters f, r and c. Chart C is used to derive the rest of the influence line through the peak value. The influence value J at any point across an influence line for a 'beam' is the fraction of the total moment on the deck, due to a load above the point, that is carried by the 'beam'. Alternatively it can be thought of as the moment (or deflection) in the 'beam' expressed as a fraction of the total moment (or deflection) that the 'beam' would experience if it carried the load by itself without load distribution to the rest of the deck. In fact the shape of the various influence lines is not uniquely defined by the peak value, and variations from the assumed geometric progression occur as shown in Fig. 10.6. However, since the sum of the values of J at all the points on a line must sum to unity, a small overestimate of J at one point is compensated for by an underestimate elsewhere. It is unlikely that design loads will be so distributed across a section

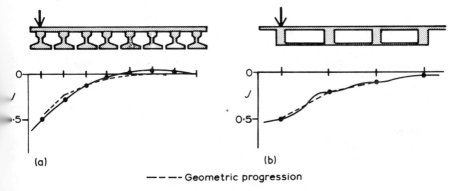

Fig. 10.6   Approximation of influence lines from charts.

that significant errors ensue. The reason the geometric decay has been assumed is that it provides a reasonable fit to the folded plate output and at the same time provides a quick method of calculation of the influence line values.

The Charts C have been called 'influence lines' as it is generally more convenient to use them as such. More strictly they are distributions of moment or deflection for the first harmonic of a line load on the relevant 'beam'. As long as the 'beams' are not very close together (so that distribution of harmonics higher than the first is not significant), the reciprocal theorem can be applied with little error to moments in the 'beams' in addition to deflections as is strictly correct. Consequently, Charts C can be used either as distributions of moment and deflection for a load on a particular 'beam' or as in influence lines.

Most design loads that are critical for longitudinal moments consist of a distributed load or a number of related point loads near midspan. The distribution of moments and deflections across the deck approximate closely to the distribution of the first harmonic as assumed in the charts. But, under the action of badly distributed loads such as a single point load or concentration of loads at one end, the distribution of moments can be worse than that of the first harmonic charts. To compensate for possible error on this count, it is generally worth taking the precaution of Morice and Little [4] of arbitrarily increasing the calculated design moments by 10 per cent.

Since the critical design load for shear force often consists of a concentration of load near one end, the distribution is worse than that of the first harmonic charts. Correction can be made as outlined in Section 12.3.3, but in general it is simpler to assume that only loads between the quarter span points are distributed while those near the ends are not distributed.

Charts C only give the values of each influence line at the 'beam' positions. The shape of the line between 'beams' depends on the rotational stiffness parameter r. If r is small as it is for decks of Fig. 10.1a and b, the deck cross-section distorts in a smooth curve. In contrast, if r is large, as for decks of

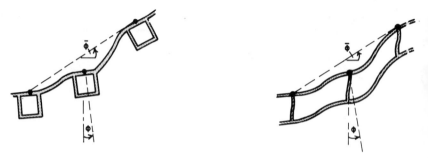

Fig. 10.7 Definition of 'beam' rotation ϕ and average rotation $\bar{\phi}$.

Charts for Preliminary Design   173

Fig. 10.1c and d, the distortion and influence lines are 'stepped'. The rotation at the 'beams' can be estimated from Charts D which give the ratio of the 'beam' rotation ϕ to the average rotation $\bar{\phi}$ calculated from the relative deflection of the 'beams' on each side, as shown in Fig. 10.7. In the case of the edge 'beam', $\bar{\phi}$ is calculated from the relative deflection of the edge and penultimate 'beams'.

10.4 Application of charts to slab deck

Fig. 10.8a and b show a solid slab deck supporting an abnormal heavy vehicle. The charts of Figs. 10.2–10.4 will be used to derive influence lines for moments at points across the cross-section, from which will be derived the moments due to the applied load.

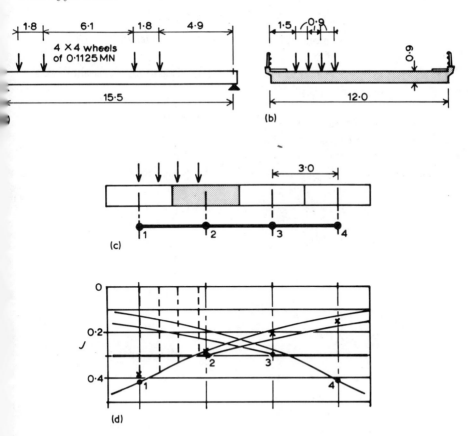

Fig. 10.8 Influence lines for slab deck. (a) Elevation (b) cross-section (c) cross-section divided into 'beams' (d) influence lines for 'beams'.

The deck is first notionally split up into identical 'beams' as shown in Fig. 10.8c. The optimum width of 'beam' for isotropic slabs is of order of ⅕ to ¼ of span.

For an isotropic slab, orthotropic slab or beam-and-slab deck the non-dimensional parameters are

$$f = 0.12 \frac{i}{l^3} \times \frac{L^4}{I} \tag{10.1}$$

$$r = 5 \frac{G}{E} \times \frac{l}{i} \times \frac{C}{L^2} \tag{10.2}$$

where

L = span
l = 'beam' spacing
i = transverse moment of inertia of slab per unit length
I = moment of inertia of 'beam'
C = torsion constant of 'beam' + l × (transverse torsion constant of slab per unit length).

If the slab has depth d and $\nu = 0.2$ so that $G/E = 0.42$,

$$f = 0.01 \frac{d^3}{l^3} \times \frac{L^4}{I} \tag{10.3}$$

$$r = 25 \frac{l}{d^3} \times \frac{C}{L^2} \tag{10.4}$$

and for an isotropic slab

$$I = \frac{ld^3}{12}$$

$$C = \frac{ld^3}{6} + l \left(\frac{d^3}{6}\right) = \frac{ld^3}{3}$$

hence

$$f = 0.12 \frac{L^4}{l^4} \tag{10.5}$$

$$r = 8.4 \frac{l^2}{L^2} . \tag{10.6}$$

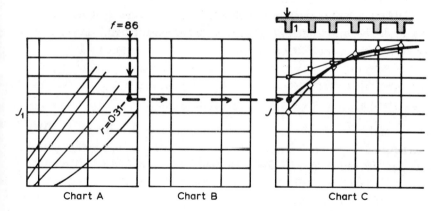

Fig. 10.9   Application of Fig. 10.2 for $f = 86$ and $r = 0.31$.

In this example, $L = 15.5, l = 3.0$, hence

$$f = 0.12 \left(\frac{15.5}{3.0}\right)^4 = 86$$

$$r = 8.4 \left(\frac{3.0}{15.5}\right)^2 = 0.31.$$

Fig. 10.9 demonstrates how Fig. 10.2 is used to derive the influence line for the edge 'beam' with $f = 86$ and $r = 0.31$. The procedure is

(1) Determine the point on Fig. 10.2, Chart A where the vertical through $f = 86$ cuts the contour of $r = 0.31$. The ordinate J_1, here equal to 0.33, is the value of the peak ordinate under the edge beam of the influence line in Chart C. Chart B is ignored for slab and beam-and-slab decks.

(2) In Fig. 10.2, Chart C interpolate the line of the influence line passing through $J_1 = 0.33$ and which lies between the lines shown for J_1 greater and less, as shown in Fig. 10.9. Hence we find the values of J at all the beams.

$$\begin{array}{cccccc} \downarrow \\ J & 0.33 & 0.22 & 0.15 & 0.10 & \Sigma = 0.80. \end{array} \qquad (10.7)$$

On a wide deck, the influence line decays to zero on the far side and the sum of all the values of J should be unity. If in addition the values of J decay in geometric progression,

$$\frac{J_n}{J_{n-1}} = (1 - J_1). \qquad (10.8)$$

It will be found that the ratio of adjacent figures in Equation 10.7 is equal to

(1–0.33). In a narrow deck such as this, the sum of all the influence values should also be unity and not 0.80 as indicated in Equation 10.7. Consequently the figures in Equation 10.7 should be corrected by scaling up by 1/0.80 to give sum of unity

$$J \quad \overset{\downarrow}{0.41} \quad 0.28 \quad 0.19 \quad 0.12 \quad \Sigma = 1.0. \tag{10.9}$$

These are the values of the influence line for edge beam 1 plotted in Fig. 10.8d.

The same procedure is followed to obtain the influence line for beam next to the edge, but using Fig. 10.3. For $f = 86$ and $r = 0.31$ we find $J_2 = 0.20$. By following the influence line in Fig. 10.3, Chart C through $J_2 = 0.20$ we obtain values of J at other beams.

$$J \quad 0.20 \quad \overset{\downarrow}{0.20} \quad 0.15 \quad 0.11 \quad \Sigma = 0.66. \tag{10.10}$$

Although J_1 must be derived from Fig. 10.3, Chart C (or with reciprocal theorem from J_2 on edge beam line) the values of J to the right of J_2 can be found using geometric reduction factor

$$\frac{J_n}{J_{n-1}} = \frac{(1 - J_1 - J_2)}{(1 - J_1)}. \tag{10.11}$$

This factor should differ very little from that of Equation 10.8. As before, the values of J in Equation 10.10 appropriate to an infinitely wide deck must be scaled up to give sum of unity, giving

$$J \quad 0.30 \quad \overset{\downarrow}{0.30} \quad 0.23 \quad 0.17 \quad \Sigma = 1.0. \tag{10.12}$$

This influence line for 'beam' 2 is also plotted in Fig. 10.8d. Also shown are the lines for 'beams' 3 and 4 which are mirror images of lines for 'beams' 1 and 2.

Although this deck has no internal 'beams' for which Fig. 10.4 is appropriate, it is worth stating here the geometric reduction factor for influence lines in Chart C to have geometric decay and sum of unity. The factor is

$$\frac{J_n}{J_{n-1}} = \frac{(1 - J)}{(1 + J)}. \tag{10.13}$$

This factor should also differ very little from that of Equation 10.8.

The total moment on the deck at the cross-section under the nearest axle to midspan due to one line of wheels is

$$M = 0.1125 \, (0.46 + 1.16 + 3.80 + 2.78)$$

$$= 0.9225 \text{ MNm}.$$

The lines of wheels coincide with points on the influence line for 'beam' 1 at which $J = 0.41, 0.37, 0.33, 0.30$. Hence, moment at 'beam' 1 due to four lines of wheels is

$$M_1 = 0.9225(0.41 + 0.37 + 0.33 + 0.30) = 1.30 \text{ MNm.}$$

Similarly,

$$M_2 = 0.9225(0.30 + 0.30 + 0.30 + 0.30) = 1.11 \text{ MNm}$$
$$M_3 = 0.9225(0.17 + 0.19 + 0.21 + 0.23) = 0.74 \text{ MNm}$$
$$M_4 = 0.9225(0.12 + 0.14 + 0.16 + 0.18) = 0.55 \text{ MNm.}$$

For design purposes, these moments are increased by 10 per cent.

Fig. 10.8d also shows by crosses the influence line for 'beam' 1 obtained from the load distribution charts of Morice and Little [4]. It can be seen that at 'beam' 1 the value obtained from the charts of this book is an over-estimate by 7 per cent, however the calculated maximum live load moment in 'beam' 1 is overestimated by only 3 per cent. When the live load moment is combined with that due to dead load, this error becomes negligible. This is because the charts were derived on the assumption that the deck is very wide. By considering only a narrow width and assuming the decay rate of influence values is the same as for a wide deck, the transverse rotation of the deck and the loading of the edge under an eccentric load are overestimated. Generally, the error has very little effect on calculated maximum design moments as it only affects narrow decks with high transverse stiffness whose load distribution is good.

10.5 Application of charts to beam-and-slab deck

10.5.1 Spaced box beams

Fig. 10.10 shows a beam-and-slab deck constructed of prestressed box beams supporting a reinforced concrete slab. The deck is supporting an abnormal heavy vehicle of four axles, each of four wheels, located at midspan between the edge beam and the next beam to the edge. The reinforced concrete has a modular ratio $m = 0.8$ relative to the prestressed concrete.

For analysis, the deck is divided into four identical 'beams' at $l = 3.2$ m spacing. For each 'beam'

$$I = 0.30 \text{ m}^4$$

$$C = 0.26 + 3.2 \left(\frac{0.8 \times 0.2^3}{6} \right) = 0.26 \text{ m}^4.$$

178 *Bridge Deck Behaviour*

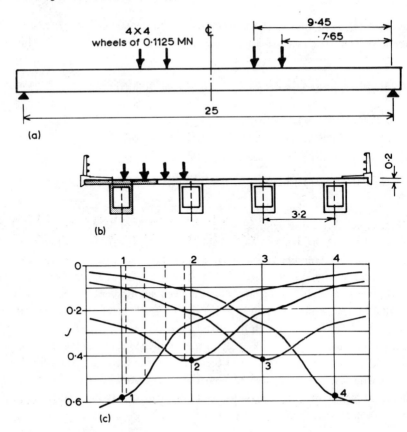

Fig. 10.10 Influence lines for beam-and-slab deck. (a) Elevation (b) cross-section (c) influence lines for 'beams'.

With $L = 25.0$ we obtain from Equations 10.3 and 10.4

$$f = 0.01 \times 0.8 \times \frac{0.2^3}{3.2} \times \frac{25^4}{0.30} = 2.5$$

$$r = 25 \times \frac{3.2}{0.8 \times 0.2^3} \times \frac{0.26}{25^2} = 5.2.$$

Using charts of Figs. 10.2–10.4 as described in Section 10.4, we find that for above values of f and r

$J_1 = 0.55$ for edge 'beam' influence line

$J_2 = 0.39$ for 'beam' next to edge influence line

$\hat{J} = 0.37$ for internal 'beam' influence line.

Charts for Preliminary Design 179

The geometric reduction factors in the influence lines are, then, from Equations 10.8, 10.11 and 10.13

$$\frac{J_n}{J_{n-1}} = (1 - 0.55) = 0.45 \qquad \text{for edge 'beam' influence line}$$

$$\frac{J_n}{J_{n-1}} = \frac{(1 - 0.25 - 0.39)}{(1 - 0.25)} = 0.48 \qquad \text{for 'beam' next to edge influence line}$$

$$\frac{J_n}{J_{n-1}} = \frac{(1 - 0.37)}{(1 + 0.37)} = 0.46 \qquad \text{for internal 'beam' influence line.}$$

Using these factors or working directly from Charts C of Figs. 10.2–10.4 we obtain influence lines for edge, next to edge (and internal 'beams' not here relevant)

$$\begin{array}{llllll} & \downarrow \\ J & 0.55 & 0.25 & 0.11 & 0.05 & \Sigma = 0.96 \end{array} \qquad (10.14)$$

$$\begin{array}{llllll} & & \downarrow \\ J & 0.25 & 0.39 & 0.19 & 0.09 & \Sigma = 0.92 \end{array} \qquad (10.15)$$

$$\begin{array}{llllll} & & & \downarrow \\ (J & 0.08 & 0.17 & 0.37 & 0.17 \ldots). \end{array} \qquad (10.16)$$

The sums of the influence values in Equations 10.14 and 10.15 are close to unity because the load distribution characteristics of the deck are not very good, and the deck is nearly wide enough for the moments in 'beams' 1 and 2 to be little influenced by loads or additional beams on the far side. On factoring up Equations 10.14 and 10.15 to give sums of unity we obtain

$$\begin{array}{llllll} & \downarrow \\ J & 0.58 & 0.26 & 0.11 & 0.05 & \Sigma = 1.0 \end{array} \qquad (10.17)$$

$$\begin{array}{llllll} & & \downarrow \\ J & 0.27 & 0.42 & 0.21 & 0.10 & \Sigma = 1.0. \end{array} \qquad (10.18)$$

These influence lines are plotted in Fig. 10.10c together with the mirror image lines appropriate to 'beams' 3 and 4.

The box beams all have high torsion stiffness so that $r = 5.2$. It can be seen in Charts D of Figs. 10.2–10.4 that for such a value of r, $\phi/\bar{\phi} = 0.5$ for all 'beams', showing that the rotation of the 'beams' is only about half of the average rotation in that region of deck. Consequently, the influence lines of Fig. 10.10c are drawn with gradients at the beam positions equal to 0.5 of the gradient of the line drawn through the 'beam' points on each side. For example, using values in Equation 10.17, we find that the rotation of 'beam' 2 is

$$\phi_2 = 0.5 \frac{(0.58 - 0.11)}{2l}. \tag{10.19}$$

In general, it is not necessary to calculate the beam rotations as ϕ can be drawn approximately by eye equal to the relevant fraction of $\bar{\phi}$.

For the abnormal heavy vehicle loading, the midspan moment per longitudinal line of wheels is

$$M = 0.1125(7.65 + 9.45) = 1.924 \text{ MNm}. \tag{10.20}$$

Since the lines of wheels coincide with values on influence lines for 'beams' 1, 2, 3 and 4 of

'beam' 1	0.57	0.49	0.35	0.27
'beam' 2	0.28	0.31	0.38	0.42
'beam' 3	0.11	0.13	0.17	0.20
'beam' 4	0.05	0.07	0.09	0.11

the total moments in the three 'beams' are

$$M_1 = 1.924(0.57 + 0.49 + 0.35 + 0.27) = 3.23 \text{ MNm}$$
$$M_2 = 1.924(0.28 + 0.31 + 0.38 + 0.42) = 2.67 \text{ MNm}$$
$$M_3 = 1.924(0.11 + 0.13 + 0.17 + 0.20) = 1.17 \text{ MNm}$$
$$M_4 = 1.924(0.05 + 0.07 + 0.09 + 0.11) = 0.62 \text{ MNm}.$$

For design, these should all be arbitrarily increased by 10 per cent.

A grillage analysis of the same deck subjected to the same load case gave moments in the four beams of 3.16 MNm, 2.52 MNm, 1.33 MNm and 0.69 MNm respectively.

10.5.2 Contiguous beam deck

A contiguous beam deck can be analysed by the charts or grillage with each analysis 'beam' representing more than one physical beam. Fig. 10.5 gives an example where each analysis 'beam' represents two physical beams. This simplification does not affect the calculated load distribution unless the summed torsional stiffness of the analysis 'beam' is large enough to cause the cross-section to distort in steps as in Fig. 10.6b. This condition can be considered satisfied if

$$r < 0.5$$

or

$$l < 0.02 \frac{d^3 L^2}{C}. \tag{10.21}$$

This limitation of beam spacing applies to slabs and beam-and-slab decks for analysis by chart or grillage.

10.6 Application of charts to cellular deck

Fig. 10.11 shows a cellular deck with thin webs and top and bottom slabs of different thicknesses. The ratio of stiffnesses of the top and bottom slabs is

$$\left(\frac{0.2}{0.15}\right)^3 = 2.4$$

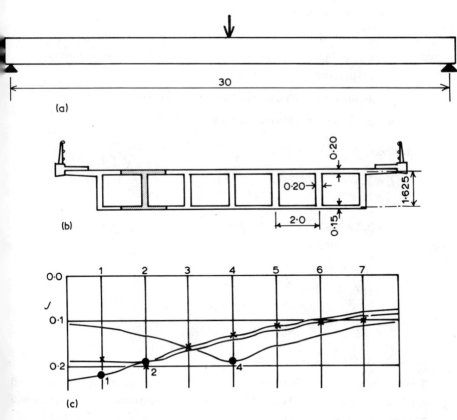

Fig. 10.11 Influence lines for cellular deck. (a) Elevation (b) cross-section (c) influence lines for 'beams'.

which is less than the limiting value for these charts stated in Section 10.3.

For a cellular deck, the non-dimensional parameters are

$$f = 0.12 \frac{i}{l^3} \times \frac{L^4}{I} \tag{10.22}$$

$$r = 6 \frac{i_w}{i} \times \frac{l}{h} \tag{10.23}$$

$$c = 0.1 \frac{G}{E} \times \frac{I_s}{I} \times \frac{L^2}{l^2} \tag{10.24}$$

where

L = span

l = 'beam' spacing = web spacing

I = moment of inertia of 'beam'

I_s = moment of inertia of slabs between webs about neutral axis of deck

i = sum of individual moments of inertia, per unit length, of top and bottom slabs

i_w = moment of inertia, per unit length, of the web

h = height of web between midplanes of slabs.

In this example,

$L = 30.0$ m $\quad l = 2.0$ m

$I = 0.51$ m^4

$$I_s = \frac{2.0 \times 1.625^2 \times 0.2 \times 0.15}{(0.2 + 0.15)} = 0.45 \text{ m}^4$$

$$i = \frac{0.2^3}{12} + \frac{0.15^3}{12} = 0.000\ 95 \text{ m}^4 \text{ m}^{-1}$$

$$i_w = \frac{0.2^3}{12} = 0.000\ 67 \text{ m}^4 \text{ m}^{-1}$$

$h = 1.625$ m.

The edge beams have slightly lower moments of inertia about the deck principal axis than internal beams. The difference here is not significant. However, as with grillage analysis, the bending stresses should be calculated using the moment of inertia assumed in load distribution analysis.

With

$$\nu = 0.15 \quad \frac{G}{E} = 0.435$$

hence

$$f = 0.12 \times \frac{0.000\ 95}{2.0^3} \times \frac{30^4}{0.51} = 23$$

$$r = 6 \times \frac{0.000\ 67}{0.000\ 95} \times \frac{2.0}{1.625} = 5.2 \quad\quad (10.25)$$

$$c = 0.1 \times 0.435 \times \frac{0.45}{0.51} \times \frac{30^2}{2^2} = 8.6.$$

The application of the charts of Figs. 10.2–10.4 using all three parameters is demonstrated in Fig. 10.12, which reproduces Fig. 10.2.

(1) Determine the point on Fig. 10.2, Chart A where the vertical through $f = 23$ cuts contour of $r = 5.2$.

(2) Move across Chart A to right edge and then move along interpolated contour line on Chart B until vertical through $c = 8.6$ is reached. The ordinate of J_1, here 0.15, is the value of the peak ordinate of the influence line at the beam in Chart C. The reduction of influence value J_1 on traversing the contour in Chart B represents the load distribution due to cellular torsion.

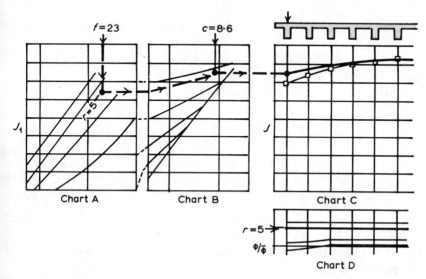

Fig. 10.12   Application of Fig. 10.2 for $f = 23$, $r = 5.2$ and $c = 8.6$.

(3) On Chart C interpolate the line of the influence line passing through $J_1 = 0.15$ which lies between the lines shown for J_1, greater and less. In this case J_1 is less than the lowest line and hence the decay away from the 'beam' must be calculated with the geometric reduction factor of Equation 10.8. Here

$$\frac{J_n}{J_{n-1}} = (1 - 0.15) = 0.85.$$

Hence influence line for edge 'beam' is

J   ↓0.15   0.128   0.108   0.092   0.078   0.066   0.057   Σ = 0.679

(10.26)

Applying Figs. 10.3 and 10.4 in similar manner we find for

'beam' next to edge $J_2 = 0.12$

internal 'beam' $\hat{J} = 0.09$.

From Equation 10.11 and 10.13, the appropriate geometric reduction factors are:

'beam' next to edge $\dfrac{J_n}{J_{n-1}} = \dfrac{(1 - 0.12 - 0.12)}{(1 - 0.12)} = 0.86$

internal 'beam' $\dfrac{J_n}{J_{n-1}} = \dfrac{(1 - 0.09)}{(1 + 0.09)} = 0.83$

with which we calculate the 'beam' values on the influence lines for 'beams' 2 and 4

J   0.120   ↓0.120   0.104   0.090   0.077   0.067   0.058   Σ = 0.636

(10.27)

J   0.052   0.063   0.075   ↓0.090   0.075   0.063   0.052   Σ = 0.47.

(10.28)

Equations 10.26–10.28 must each be scaled up to give the corrected influence line with sum of unity.

				↓				
J	0.22	0.19	0.16	0.14	0.11	0.10	0.08	Σ = 1.0

			↓					
J	0.19	0.19	0.16	0.14	0.12	0.11	0.09	Σ = 1.0

				↓				
J	0.11	0.13	0.16	0.19	0.16	0.13	0.11	Σ = 1.0.

These influence lines are plotted in Fig. 10.11c. Since the factor r is equal to 5.2, Charts D of Figs. 10.2–10.4 indicate that $\phi/\bar{\phi} = 0.5$ as in the last example. Hence influence lines are stepped, as in last example, with rotations at 'beams' approximately equal to half the average rotation between 'beams' on two sides.

The influence lines are used for the calculation of moments in the same manner as demonstrated in the previous two examples.

The deck of this example has proportions which differ only slightly from those of a perspex model reported by Sawko [10]. The model, of span 36.8 in, had six cells of inside dimensions 2 × 2 in with all webs and slabs 0.25 in thick. For these dimensions, $f = 31$, $r = 3$, $c = 8.6$; it will be found in the charts that these non-dimensional parameters give the same influence lines and deflection distributions as the example above. Consequently, the lines of Fig. 10.11c can also be considered as chart predictions for the deflection profiles of the model. Sawko reported the predictions of finite element analyses; the deflection profile for a load above the web next to the edge is shown in Fig. 10.11c by crosses. It is evident that the chart prediction is very close.

REFERENCES

1. Pucher A. (1964), *Influence Surfaces of Elastic Plates*, Springer Verlag, Wien and New York.
2. Rusch, H. and Hergenroder, A. (1961), *Influence Surfaces for Moments in Skew Slabs*. Technological University, Munich. Translated from German by C. R. Amerongen, Linden, Cement and Concrete Association.
3. Balas, J. and Hanuska, A. (1964), *Influence Surfaces of Skew Plates*, Vydaratelstro Slovenskej Akademic Vied, Bratislava.
4. Morice, P. B. and Little, G. (1956), *Right Bridge Decks Subjected to Abnormal Loads*, Cement and Concrete Association, London.
5. Rowe, R. E. (1962), *Concrete Bridge Design*, C. R. Books, London.
6. Cusens, A. R. and Pama R. P. (1971), 'Design curves for the approximate determination of bending moments in orthotropic bridge decks,' Civil Engineering Department, University of Dundee.
7. *Idem*, (1970), 'Design curves for the approximate determination of twisting moment and longitudinal shear in orthotropic bridge decks,' Civil Engineering Department, University of Dundee.
8. Department of the Environment (1973), 'Bridge Deck Design Charts for

Orthotrophic right bridge decks,' Highway Engineering Computer Branch HECB/B1/5.
9. Department of the Environment (1970), Ministry of Transport Technical Memorandum: Shear Key Decks, Annexe to Technical Memorandum (Bridges), No BE 23.
10. Sawko, F. (1968), 'Recent developments in the analysis of steel bridges using electronic computers,' BCSA Conference on Steel Bridges, London, June, pp. 39–48.

11
Temperature and prestress loading

11.1 Introduction

This chapter describes the actions of temperature loading and prestress on bridge decks, and shows how these loads can be simulated in the analytical methods outlined in previous chapters. Although temperature loading and prestress are described below one after the other, it is important to emphasize the very different principles of these two types of loading.

11.2 Temperature strains and stresses in simply supported span

The rise of temperature in an element of material causes the element to expand if it is unrestrained. Alternatively, if the element is prevented from expanding, the rise in temperature causes an increase in stress which depends on Young's Modulus of the material. Either the increase in strain or the increase in stress can be taken as the starting point for the calculation of the distribution of temperature stresses in the structure. For a simply supported deck with linear variation in temperature between top and bottom surfaces, the simplest way to calculate the flexure is to work from the temperature-induced free strain. However, for the more general bridge problems with complicated deck geometry and non-linear temperature distributions, it is simplest to start from the assumption that the deck experiences temperature changes which set up stresses

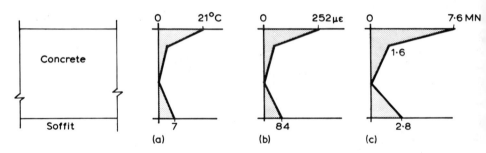

Fig. 11.1 Temperature effects in slab deck. (a) Temperature distribution (b) strains in unrestrained slice   (c) stresses in restrained slice.

while the deck is rigidly restrained throughout, and then calculate the effects of removing the theoretical restraints.

Fig. 11.1a shows an element of deck with non-linear temperature distribution varying from +21°C at the top surface through 0 to +7°C at the soffit. (The assumed temperature distribution relevant to a particular bridge is derived from a heat flow calculation or from the appropriate code of practice.)

If the coefficient of thermal expansion is α, the unrestrained thermal strains are

$$\text{expansion} \quad \epsilon = \alpha \Delta T. \tag{11.1}$$

If $\alpha = 12 \times 10^{-6} \ °C^{-1}$, the temperatures of Fig. 11.1a cause the strains shown in (b) in an unrestrained thin slice of the deck. With such strains, plane sections do not remain plane.

When expansion is prevented, with plane sections held plane, then the locked-in stresses are

$$\text{compression} \quad \sigma = \alpha E \Delta T \tag{11.2}$$

where E is Young's Modulus.

For $E = 30\ 000 \ \text{MNm}^{-2}$, the temperatures of Fig. 11.1a cause the locked-in stresses of (c) in a rigidly restrained slice of deck.

It is important to note that the strains of Fig. 11.1b and stresses of (c) are alternative primary effects of temperature change dependent on boundary conditions. Primary stress and strain are not proportional in accordance with Hook's Law, as stress is large when strain is kept small and vice-versa. The temperature has in effect induced a 'lack of fit' or a 'jacked-in force'.

In contrast, the secondary stresses and strains due to redistribution of primary moments are related by Hook's Law if the material is elastic.

The locked-in stress distribution of Fig. 11.1c (shown again as Fig. 11.2a) can

Temperature and Prestress Loading

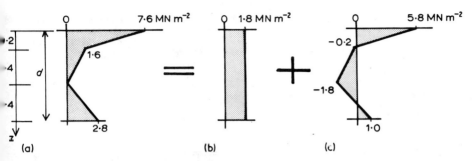

Fig. 11.2 Temperature stresses in restrained deck. (a) Stresses in restrained slice (same as Fig. 11.1c) (b) compression P (c) restrained moment M_T and residual.

be thought of as composed of two parts which affect the structure in different ways:

(1) an average compression stress shown in Fig. 11.2b causing resultant force on cross-section of P

(2) a non-linear stress distribution, shown in Fig. 11.2c which comprises a moment M_T and residual stresses with no net resultants.

For the particular stresses of Fig. 11.2a, the average compression stress of (b) is simply

$$\bar{\sigma} = \int_0^d \frac{\sigma b \, dz}{\text{Area}} \tag{11.3}$$

where b = breadth.

Here

$$\bar{\sigma} = \frac{1}{1.0} \left[\frac{(7.6 + 1.6)}{2} 0.2 + \frac{(1.6 + 0)}{2} 0.4 + \frac{(0 + 2.8)}{2} 0.4 \right]$$

$$= 1.8 \text{ MN m}^{-2}.$$

Fig. 11.2c is simply (a) minus this average stress in (b).

When a length of deck is rigidly restrained as in Fig. 11.3a, the stresses of Fig. 11.2a act on every cross-section. On internal cross-sections the stresses on two sides balance, so that it is only at the end faces that these stresses must be balanced by the externally applied restraining forces. When longitudinal expansion restraint is removed, as in Fig. 11.3b, the deck expands and the compression stresses of Fig. 11.2b are relaxed. Only the moment and residual stresses of Fig. 11.2c remain. When the moment restraint is removed as in Fig. 11.3c, the deck flexes. In effect, the restraining moments have been

190 *Bridge Deck Behaviour*

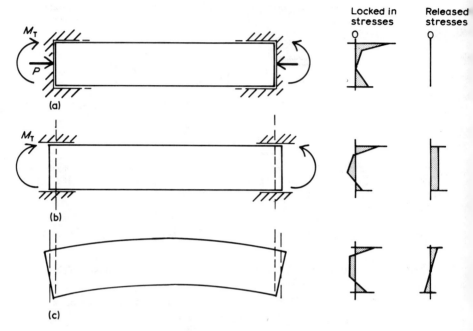

Fig. 11.3 Progressive release of temperature forces. (a) Fully restrained (b) compression release   (c) flexure release.

cancelled by equal and opposite relaxing moments which cause flexure. The magnitudes of the restraining (and opposite relaxing) moments are equal to the moment of the stress diagram of Fig. 11.2c (about any point).

$$M = \int_0^d \sigma b z \, dz. \tag{11.4}$$

Here

$$M = \frac{5.8}{2} \times 0.193 \times 0.064 - \frac{0.2}{2} \times 0.007 \times 0.198 - \frac{(0.2 + 1.8)}{2} \times 0.4 \times 0.45$$

$$- \frac{1.8}{2} \times 0.26 \times 0.69 + \frac{1.0}{2} \times 0.14 \times 0.95$$

$$= -0.27 \text{ Mnm m}^{-1} \text{ width.} \tag{11.5}$$

The stresses and strains caused by the relaxing moment are related by Hook's Law and exhibit the elastic behaviour described in Chapters 2 and 3. In particular, the stress distribution is linear as shown in Fig. 11.4b with stress

Temperature and Prestress Loading 191

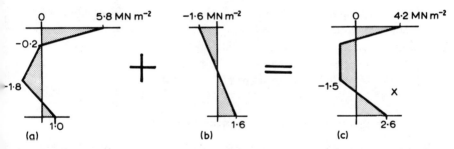

Fig. 11.4 Temperature moment and residual stresses. (a) Restrained moment M_T and residual (same as Fig. 11.2c) (b) relaxing moment M_T (c) residual.

related to M by

$$\sigma = \frac{M}{Z}. \tag{11.6}$$

For a rectangular beam $Z = bd^2/6$, and hence the stress relaxation at surfaces in Fig. 11.4b due to $M = -0.27$ is

$$\sigma = -\frac{0.27}{1^2/6} = -1.6 \text{ MN m}^{-2}.$$

When the relaxing stress distribution of (b) is added to the restrained stresses of (a), we obtain the residual stress distribution of (c). This has no net compression or moment, as can be checked by reapplying Equations 11.3 and 11.4.

The residual stress diagram Fig. 11.4c shows the final distribution of temperature-induced stresses on a cross-section of a simply supported span. The maximum stresses are large, even though the deck is simply supported, because the temperature distribution is so non-linear. It is only when the temperature distribution is linear that its induced stresses are linear in (a) so that when relaxed by an equal and opposite linear stress distribution of (b), no residual stresses (c) remain on the section.

At the end faces of the deck, the residual temperature stresses are not resisted by external forces. Consequently, these forces redistribute by local elastic distortion over a length of deck approximately equal to the depth of the section. This redistribution is accompanied by local high longitudinal shear forces which transfer the residual compression forces near the top and bottom faces to the opposed residual tension in the middle. In this example, the residual compression force below level X in Fig. 11.4c is

$$\frac{2.6}{2} \times 0.25 = 0.33 \text{ MN m}^{-1} \text{ width.}$$

Thus the longitudinal shear force at level X near the end of the deck is also 0.33 MN m^{-1} width.

In the above example it is assumed that the deck has the same breadth at all levels. If the deck is made up of beams with breadth b varying with depth, the variation in b is included in Equations 11.3 and 11.4

11.3 Temperature stresses in a continuous deck

The relaxation of temperature stresses in a continuous deck is only slightly more complicated than in a simply supported span. Fig. 11.5a shows a three span deck which is fully restrained with a locked-in non-linear temperature distribution.

The compression force due to average temperature compression stress of Fig. 11.2b is relaxed by letting all the spans expand freely in the longitudinal direction. Fig. 11.5b shows the restrained temperature moments on each span similar to Fig. 11.3. Since the moments on the two sides of the internal supports balance, it is possible to join the spans together as in (c) without restraints at internal supports and without any relaxation taking place. In other words, only moments at the ends of the deck are needed to restrain a continuous deck against temperature flexure. Relaxation of these end restraining moments is achieved by superposing equal and opposite relaxing moments as shown in Fig. 11.5d. In contrast to a simply supported span which hogs with uniform moment when subjected to end relaxing moments, the continuous deck distributes the relaxing moments in accordance with the continuous beam theory of Chapter 2 as shown in Fig. 11.5d. Consequently, superposition of the stresses from the restrained temperature moments of (c), which are uniform along the deck, and the varying relaxing moments of (d) give different stress distributions at different points along the deck.

At the ends of the end spans, the relaxing moment is equal and opposite to the restrained temperature moment, and the stress distribution is similar to Fig. 11.4c. However at the next support, the 'relaxing moment' in Fig. 11.5d has an opposite sign to that at the ends and so does not counter-balance the restrained temperature moment and in fact increases the combined top stresses as shown in Fig. 11.6.

If the deck has a section that varies along its length or if the temperature distribution changes along the deck, the restrained temperature moment also changes along the deck. Fig. 11.7 shows two alternative methods of representing a haunched deck. If the deck is represented by a number of connected uniform segments of different section as on the left side of (b), the temperature flexure will be restrained by different moments in each segment. When the segments are

Temperature and Prestress Loading 193

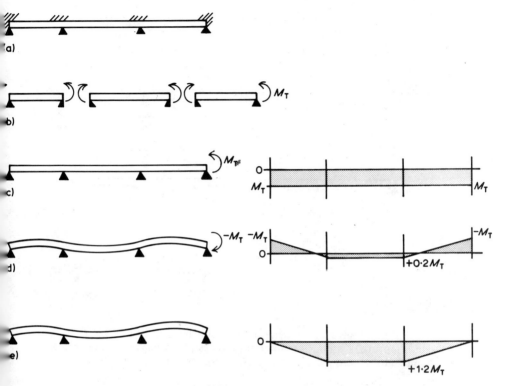

Fig. 11.5 Release of temperature moments in continuous deck. (a) Restrained deck (b) restrained moments in each span (c) restrained moments in connected spans (d) relaxing flexural moments (e) final moments = c + d.

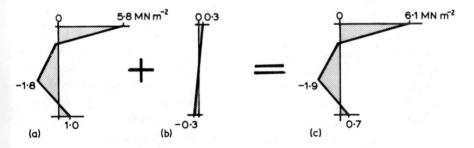

Fig. 11.6 Temperature stresses at internal support of continuous deck of Fig. 11.5. (a) Restrained moment M_T and residual (same as Fig. 11.2c) (b) relaxing moment = $0.2\,M_T$ (c) total stresses.

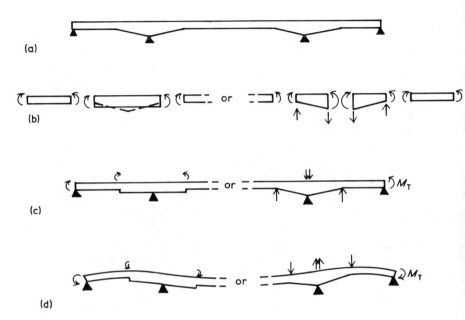

Fig. 11.7 Release of temperature moments in haunched deck. (a) Deck elevation (b) restrained temperature forces on each span (c) restrained temperature forces on connected spans (d) relaxing flexural forces.

connected as in (c), the moments do not balance at internal connections. The relaxing moments applied to the continuous deck must cancel the external restrained temperature moments, and so out-of-balance moments must be applied at changes of section of continuous beam as in (c). Alternatively, if a segment is tapered, the restrained temperature moments are different on its two ends. To maintain equilibrium, opposed vertical forces must be applied to the two ends of the segment, as on the right of (b). (This shear force is statically equivalent to a uniformly distributed moment applied along the taper.) When the segments are connected, the restrained temperature moments now balance at the connections but the vertical restraining forces remain. As a consequence, the cancelling relaxing forces applied to the continuous beam for distribution include vertical forces at the ends of tapers as well as moments at the end supports.

Wide bridges experience stresses due to the thermal expansion of an element being resisted transversely as well as longitudinally. The behaviour is the same in the two directions. However if Poisson's Ratio is significant, it may be necessary to investigate the interaction of stresses and flexure in the two directions. This is complicated, and outside the scope of this chapter.

11.4 Grillage analysis of temperature moments

The analysis of distribution of temperature moments in a two-dimensional deck by grillage is the same in principle as that for continuous beam in Section 11.3. Fig. 11.8 shows a grillage for a two span skew deck. The restrained temperature moments and cancelling flexural moments on every elemental beam are similar to those on the elements of Fig. 11.5b. When the elements are connected, only the out-of-balance moments (and forces) remain applied to the joints. In this example it is assumed that the slab deck is uniform so that the moments on the element beams on the two sides of the internal joints cancel. At the edges, the temperature-induced stresses of Fig. 11.4a must be resisted by equal and opposite external forces while flexure is restrained. When these moments are cancelled by relaxing moments, they are only applied to the edges of the deck as in Fig. 11.8. The relaxing moments distribute non-uniformly throughout the deck as in the continuous beam in Fig. 11.5b, and the precise distribution must be found from the grillage analysis. The final stresses at any point are simply the superposition of the stresses from the grillage output moments and the restrained temperature stresses of Fig. 11.4a.

In wide multicell decks with few diaphragms, the top and bottom slabs at points far from a diaphragm can expand and contract with little resistance from the deck except the out-of-plane flexure of the webs. A three-dimensional analysis is necessary if a detailed study of temperature stresses is to be made of such a deck or of any other complex structure. Most three-dimensional structural analysis programs have the facility to input directly different

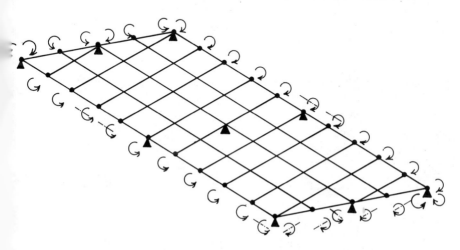

Fig. 11.8 Relaxing flexural moments applied to grillage.

temperatures in members without the user having to calculate equivalent loads or lack of fit.

11.5 Differential creep and shrinkage

The physical action of differential creep or shrinkage in a structure is similar to that of temperature. If a thin slice of deck is unrestrained, differential creep and shrinkage will cause non-uniform strains in the slice in the same way as temperature in Fig. 11.1b. The only difference is that the creep and shrinkage strain diagram is usually stepped. Also, if the slice is restrained rigidly, the differential creep and shrinkage usually induce a tensile stress distribution (similar to temperature compression Fig. 11.1c) with tensile stresses equal to Young's Modulus x the free creep and shrinkage strains. The distribution of secondary moments is calculated in the same manner as for temperature moments.

11.6 Prestress axial compression

A prestressing cable in a concrete bridge deck subjects the concrete to three different systems of loading, illustrated in Fig. 11.9.

(1) Axial compression in concrete due to compression force applied to concrete by anchorage of cable or friction forces, shown in Fig. 11.9a;

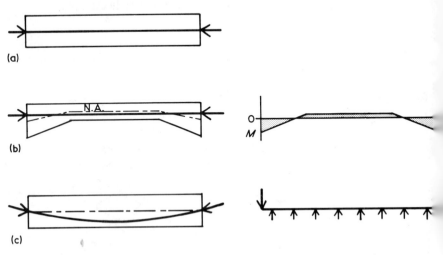

Fig. 11.9   Forces on concrete due to prestress. (a) Axial compression (b) moments due to eccentricity of compression   (c) vertical (or horizontal) forces due to cable curvature.

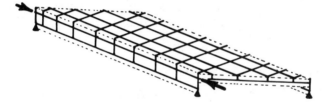

Fig. 11.10 Cruciform space frame for analysis of prestress force field.

(2) moments due to eccentricity of resultant compression force from neutral axis of concrete section, shown in (b);

(3) vertical loads on concrete due to reaction of curved cable, shown in (c).

This and subsequent sections give a brief description of how the concrete structure behaves under these load systems, and then demonstrate how the loads are simulated for application to the analytical models described in preceding chapters. The general behaviour of prestressed concrete and its design is described much more thoroughly in references [1 and 2].

The prestress compression force on a long beam subjects cross-sections from end to end to a uniform stress equal to the force divided by the section area. The distribution of stress is only complicated in the region of the anchorage and ends of flanges where local shear lag action is significant. The distribution of prestress compression stress in a wide deck can also be simple if the prestress anchorages are evenly distributed across the section, or if the density of prestress varies linearly from one side to the other. In the latter case the deck is bent in plan, and complications only arise if the supports are rigid against the small sideways deflections. However, if the deck has a complicated shape in plan, if the prestress cables are curved in plan, or if one part is stressed more (or before) the rest, some form of plate analysis may be necessary. This could be a space frame (described in Chapter 7) or a finite element analysis (described in Chapter 13). Fig. 11.10 shows a cruciform space frame that can be used to study the distribution of forces near the obtuse corner of a skew beam-and-slab deck.

11.7 Prestress moments due to cable eccentricity

Fig. 11.11a shows a continuous deck with constant cross-section subjected to prestress by a straight cable with force P at eccentricity e from the neutral axis. Because the concrete only comes into contact and reacts with the cable at the end anchorages, the only forces applied to the concrete are compression P and hogging moment Pe at anchorages. Since the neutral axis is straight, the compression force P simply subjects the whole length of deck to uniform

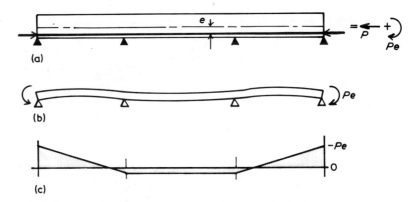

Fig. 11.11 Prestress moments in three span deck due to eccentricity of prestress. (a) Elevation (b) moments applied to concrete (c) moment diagram.

compression, which does not affect continuous beam analysis. The end moments make the deck flex, but because it is held down by internal supports, the moments distribute non-uniformly as shown in Fig. 11.11c. This distribution is found from continuous beam analysis with beam subjected to end moments as in (b).

When the deck has a varying section as in Fig. 11.12a, the changing eccentricity of the axial force P subjects the concrete to varying applied moment along its length. The effect can be visualized by considering the beam (or grillage) restrained at the support positions and at the changes of section. The fixed end moments Pe are applied to each end of each section, and a continuous beam or grillage analysis is then used to determine the redistribution on release and removal of the restraints. At internal changes of section the fixed end moments Pe on each side largely cancel, and only the difference ΔPe need be applied. In tapers, as on right of Fig. 11.12, Pe is the same on both sides of any section and so cancels out. However, the concrete along the taper is subjected to moment ΔPe which must be applied uniformly along the taper, or, as is much simpler, this uniformly distributed applied moment is represented by statically equivalent vertical loads $\Delta Pe/l$ on the two ends of the taper as shown in Fig. 11.12c and d. Changes ΔPe of prestress moment along the deck due to decrease of P by friction can also be represented in the beam or grillage analysis by opposed vertical loads $\Delta Pe/l$ at the two ends of the frictional change.

11.8 Prestress moments due to cable curvature

A curved prestress cable presses against the concrete on the inside of the curve and thus subjects the deck to vertical loads. The cable profile is usually chosen

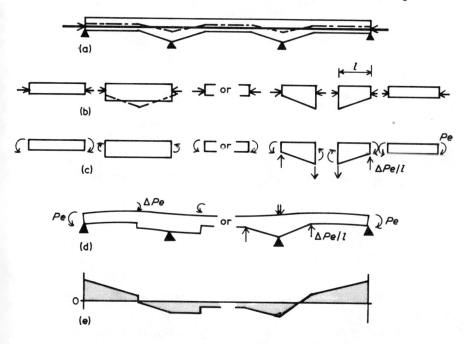

Fig. 11.12 Prestress moments due to eccentricity in deck of varying section. (a) Elevation (b) compression on restrained elements (c) moments due to eccentricity of compression on restrained elements (d) moments and forces applied in continuous beam analysis (e) diagram of distributed moments.

so that these vertical loads on the concrete largely cancel out the loads and bending stresses in concrete due to self weight. Since the vertical prestress forces press on the concrete in the same way as live or dead loads, their effects can be investigated with continuous beam or grillage in precisely the same way.

Fig. 11.13a shows the forces acting on a short length of concrete due to the curvature of the cable in (b). If the inclination of the cable changes from θ_1 to θ_2, the vertical force is

$$\text{Vertical force} = P(\sin\theta_2 - \sin\theta_1). \tag{11.7}$$

If the cable curve has a prabolic shape, the load is uniformly distributed with

$$\text{u.d.l.} = P(\sin\theta_2 - \sin\theta_1)\frac{1}{l}. \tag{11.8}$$

Strictly, the vertical loads due to prestress should be applied to the continuous beam or grillage analysis as distributed loads with intensity varying with curvature. But if the beam is notionally split up into short elements as in a grillage, the prestress vertical loads can be applied as point loads at the joints

200   Bridge Deck Behaviour

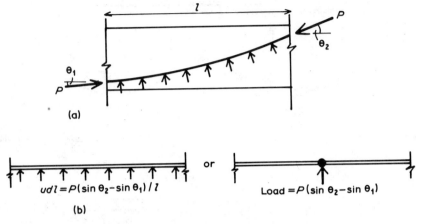

Fig. 11.13   Prestress loading due to cable curvature. (a) Forces on concrete (b) loading for continuous beam or grillage analysis.

with θ_1 and θ_2 in Equation 11.7 being the cable gradients midway along the elements to each side.

Fig. 11.14 summarizes the loading that has to be applied to a continuous beam or grillage to simulate the effects of anchorage eccentricity, varying section and cable curvature. Working from left to right the loads are:

(1) Hogging moment Pe_1 simulates moment on concrete at anchorage due to eccentricity e_1 of anchorage below neutral axis.

(2) Vertical force $P \sin \theta_1$ simulates vertical component of force on anchorage due to inclination θ_1 of cable.

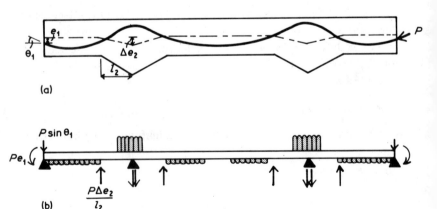

Fig. 11.14   Example of prestress loading for continuous beam analysis. (a) Cable profile   (b) loading.

(3) Distributed upwards load along length of deck that cable is curving upwards.

(4) Vertical forces $P\Delta e_2/l$ at each end of haunch to simulate bending loads on deck due to change in level Δe_2 of neutral axis along haunch.

(5) Distributed downwards load above support along length of deck that cable is curving downwards.

It should be noted that the sum of all vertical loads on the deck due to prestress should be zero.

11.9 Prestress analysis by flexibility coefficients

The analysis of prestress moments in Sections 11.7 and 11.8 is orientated towards a grillage analysis in which the structure is in effect held rigidly at the joints, subjected to the vertical loads and fixed end moments due to prestress on the elements, and released to redistribute the moments. An alternative approach can be used in which the structure is first considered as simply supported spans subjected to the moments due to prestress, and then the reactant moments are found that are necessary to make the segmented structure join together. This is the flexibility coefficient method described in Section 2.3.5.

Fig. 11.15 shows a slice of deck. The prestress cable crosses the two faces of the slice with tensions P_1 and P_2 and eccentricities e_1 and e_2. When the deck is simply supported, the compression P in the concrete (shown in Fig. 11.15) at any section is coincident and opposite to the cable tension. The free moment in

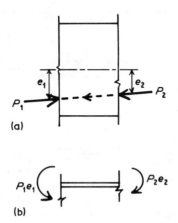

Fig. 11.15 Prestress forces on free element of deck. (a) Forces (b) equivalent moments.

the concrete at that point is simply

$$M = Pe. \qquad (11.9)$$

where P is the prestress force at the section and e is the eccentricity of the prestress cable from the neutral axis. If the cable is curved and the level of the neutral axis moves up and down as in Fig. 11.16a, e is still the distance between these two lines. (b) shows the free moment diagram along the spans. It is similar

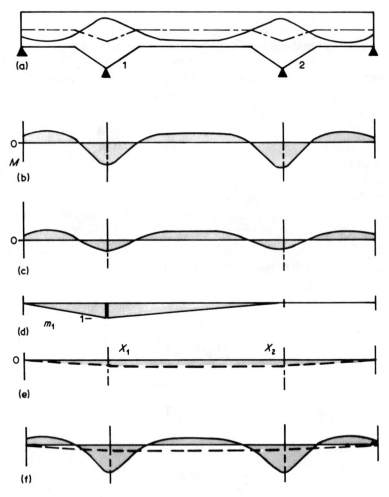

Fig. 11.16 Flexibility analysis of prestress moments in continuous beam. (a) Cable profile (b) free moment $M = Pe$ diagram (c) free M/EI diagram (d) moments for unit release at 1 (e) reactant, parasitic (or secondary) moments (f) total prestress = b + e.

in shape to the plot of the cable off the neutral axis, except that the moment is reduced away from the anchorages due to the drop in P by friction. Under the action of these free moments, the various spans will deflect and, since they are at present considered as separate spans, relative rotations occur between adjacent ends of spans on each side of the internal supports. At support 1 the relative rotation is

$$\delta_{1W} = \int \frac{m_1 m_W \, ds}{EI} \tag{2.12}$$

where m_W is the free moment Pe due to applied prestress and m_1 is the moment due to the unit reactant moment at support 1, shown in Fig. 11.16d. To calculate product integral (Equation 2.12) it is necessary to plot Pe/EI as in (c) by dividing moment diagram of (b) by local EI. The product integral can then be computed by multiplying (c) by m_1 in (d), thus giving δ_{1W}. The relative rotation δ_{2W} at support 2 is found in similar manner. By following the procedure outlined in Section 2.3.5, δ_{12} (and δ_{21}) is found, but with change of EI considered. Equations of the form of Equation 2.11 are then obtained to give reactant moments X_1 and X_2 at the supports in Fig. 11.16e. These moments are added to the free moments of (b) to give the total moments in the continuous concrete structure as shown in (f).

It should be noted that for simply supported spans, prestress is simply an interaction between cable and concrete and there are no external reactions. When the spans are made continuous the reactant moments X_1, X_2, etc. induce reactions at the support. Often the free moments are referred to as 'primary prestress' and the reactant moments as 'secondary' or 'parasitic' moments. This distinction is convenient for continuous beams, but it has little meaning in wide two-dimensional decks which are skew, tapered or curved since the lateral redistribution of moments due to the deck itself is as significant as longitudinal redistribution due to continuity at supports.

REFERENCES

1. Libby, J. R. (1971), *Modern Prestressed Concrete: Design Principles and Construction Methods*, Van Nostrand Reinhold, New York.
2. Harris, J. D. and Smith, I. C. (1963), *Basic Design and Construction in Prestressed Concrete*, Chatto and Windus, London.

12
Harmonic analysis and folded plate theory

12.1 Introduction

The grillage and finite element methods described in other chapters analyse complicated bridge decks by considering them assemblages of simple structural elements, for each of which simplified force-deflection behaviour is assumed. Numerous discontinuities of structure are introduced to make solution possible. One method of analysis which does not require the structure to be cut up in the same way uses harmonic analysis and either 'orthotropic' or 'folded' plate theory. Exact solutions are possible in the sense that the mathematic model can be made to rigorously satisfy the stress-strain assumptions for an elastic continuum. However the methods, as described below, can only be used for prismatic decks whose cross-section is the same from abutment to abutment. Fig. 12.1. shows examples of cross-sections of common types of deck which do have such sections. (a) and (b) can be analysed by orthotropic plate theory, while (c), (d) and (e) can be analysed by folded plate theory.

In harmonic analysis, the load is broken down into a number of components, each consisting of a distributed line load parallel to the structure and with intensity varying as a pure sinewave as in Fig. 12.2. Under the action of each sinewave component, every longitudinal slab or web of the structure deflects and twists in a pure sinewave. Since every differential of a sinewave is a sine or cosine function, the equilibrium equations, which can be thought of as differentials of

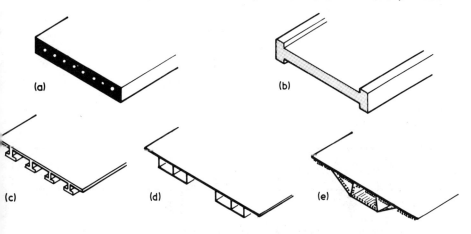

Fig. 12.1 Cross-sections of prismatic decks.

deflections, can also be expressed as a number of sine or cosine functions without differential symbols. These equations can be solved as conventional simultaneous equations.

This chapter explains how a load or other function can be represented as a number of harmonic components. It is then shown how the load distribution characteristics of a deck are markedly different for low and high harmonics so that the severity of different types of load can be anticipated from their harmonic composition without detailed analysis of the structure. Finally, the basic principles of folded plate theory are summarized.

12.2 Harmonic components of load, moment, etc.

The general theory of Harmonic, or Fourier, Analysis is described in detail in many mathematics text books and handbooks, including Kreyszig [1]. This section describes only a particular application of the theory to deck analysis.

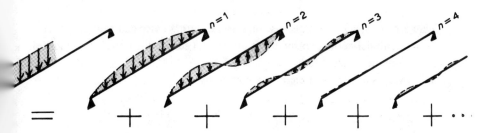

Fig. 12.2 Breakdown of load into sinusoidal harmonic components.

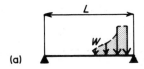

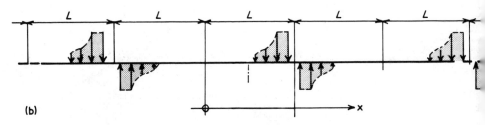

Fig. 12.3 (a) Simple span as part of (b) infinite beam.

Fig. 12.3a shows a simply supported beam of span L with coordinate origin at left support. The beam supports a load. It is possible to define a generalized infinite series of harmonic components for the loading, and then calculate the magnitudes of the harmonics relevant to the boundary conditions of zero moment and zero deflection at support positions. However, it is simpler to satisfy the boundary conditions using physical reasoning. The beam and load can be considered as part of an infinitely long beam with repetitive reversed loading as in (b). Automatically there are points of zero moment and deflection at points of contraflexure spaced regularly L apart. The repetitive reversed loading W, which is a function of x, can be represented by an infinite series of sine functions of form

$$W(x) = b_1 \sin\frac{\pi x}{L} + b_2 \sin\frac{2\pi x}{L} + \cdots + b_n \sin\frac{n\pi x}{L} + \cdots \qquad (12.1)$$

There are no cosine terms because on differentiating twice and four times for moment and deflection, cosine functions do not satisfy the boundary conditions of no moment or deflection at $x = 0$ and L.

The values of the coefficients b are given by

$$b_n = 2 \times \left[\text{average value of } W(x) \sin\frac{n\pi x}{L} \text{ between } x = 0 \text{ and } x = L\right]. \quad (12.2)$$

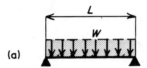

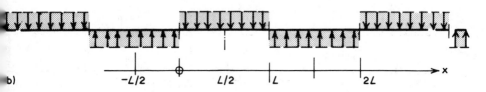

Fig. 12.4 Uniformly distributed load on simple span.

For example, for the case of the uniformly distributed load shown in Fig. 12.4a,

$$b_n = 2 \times \frac{1}{L} \int_0^L \left(\frac{W}{L}\right) \sin\frac{n\pi x}{L} = 2 \times \frac{1}{L} \times \frac{W}{L} \times \frac{2L}{n\pi} \quad \text{for } n = 1, 3, 5, \ldots \quad (12.3)$$

$$= 0 \quad \text{for } n = 2, 4, 6, \ldots$$

All the coefficients for even number harmonics in the above example are zero because the uniformly distributed load is symmetric about the deck centre line, whereas even number harmonic functions are all antisymmetric. It is possible to represent any load case, such as the point load in Fig. 12.5, as a combination of a symmetric component shown in (b), and an antisymmetric component shown in (c). The harmonic representation of the symmetric component is composed entirely of odd number harmonic functions, while the antisymmetric component is represented entirely by even number harmonics.

Thus for load, moment or deflection:

Symmetric component harmonics $\quad n = 1, 3, 5 \ldots$

Antisymmetric component harmonics $n = 2, 4, 6 \ldots$ (12.4)

The shear force diagrams for the load case of Fig. 12.5a and its symmetric and antisymmetric components in (b) and (c) are shown in Fig. 12.6a, b and c. These should also be thought of as parts of infinitely long repetitive diagrams. It

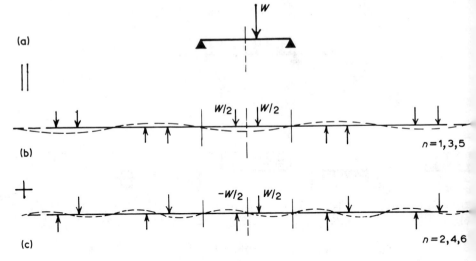

Fig. 12.5 Symmetric and antisymmetric components of a load. (a) Total (b) symmetric (c) antisymmetric.

should be noted that the symmetric load produces an antisymmetric shear force diagram and vice versa.

The general expression for the shear force diagram can be written

$$S(x) = a_1 \cos\frac{\pi x}{L} + a_2 \cos\frac{2\pi x}{L} + \cdots + a_n \cos\frac{n\pi x}{L} + \cdots \qquad (12.5)$$

where

$$a_n = 2 \times [\text{average value of } S(x) \cos\frac{n\pi x}{L} \text{ between } x = 0 \text{ and } x = L]. \quad (12.6)$$

It will be found for shear force and slope that:

Symmetric component harmonics $\quad n = 2, 4, 6 \ldots$

$$(12.7)$$

Antisymmetric component harmonics $n = 1, 3, 5 \ldots$

In practice, it is generally easier to remember that shear force is the integral of load intensity times -1, simply to integrate the load function. Thus if

$$W = \Sigma\, b_n \sin\frac{n\pi x}{L} \qquad n = 1, 2, 3 \ldots$$

$$(12.8a)$$

$$S = \Sigma\, b_n \left(\frac{L}{n\pi}\right) \cos\frac{n\pi x}{L}$$

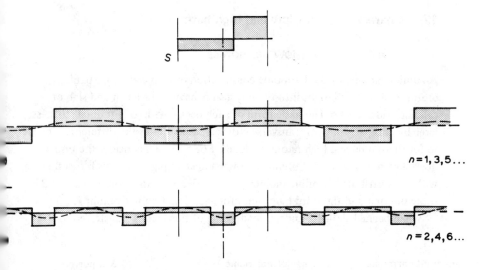

Fig. 12.6 Shear force diagram for loads of Fig. 12.5. (a) Total (b) antisymmetric (c) symmetric.

and similarly, for bending moment, slope and deflection we obtain from repeated integration

$$M = \Sigma\, b_n \left(\frac{L}{n\pi}\right)^2 \sin\frac{n\pi x}{L}$$

$$EI\frac{dw}{dx} = \Sigma\, b_n \left(\frac{L}{n\pi}\right)^3 \cos\frac{n\pi x}{L} \qquad (12.8b)$$

$$EIw = \Sigma\, b_n \left(\frac{L}{n\pi}\right)^4 \sin\frac{n\pi x}{L}.$$

It is useful to remember that, for any harmonic, the coefficients of the load, bending moment and deflection are proportional in the ratio

$$1 : \left(\frac{L}{n\pi}\right)^2 : \left(\frac{L}{n\pi}\right)^4 \times \frac{1}{EI}.$$

The amplitudes of harmonic components for typical design loads are tabulated in Figs. A.2 and A.3. Column 1 of Fig. A.3 shows the total load function with its integrals for shear force, etc., derived from simple beam theory. Column 2 gives the magnitude of the first harmonic. Column 3 gives the sum of all higher harmonics, which is simply the difference between columns 1 and 2. Column 4 gives the amplitude of any other harmonic n.

12.3 Characteristics of low and high harmonics

12.3.1 Distribution of low harmonics

An understanding of the harmonic composition of load can be particularly useful in obtaining insight into the physical behaviour of beam-and-slab or cellular bridge decks. The basic form of such decks, with longitudinal stiffness large by comparison with transverse stiffness, accentuates the different response to low harmonic and high harmonic loads. For example, consider the bridge deck shown in Fig. 12.7. Column 1 shows a load acting on one web together with the distributed bending moments and deflected form. Columns 2 and 3 show the first and the third harmonics for each function in Column 1.

When beam 1 deflects under a load, some of this load is transferred to

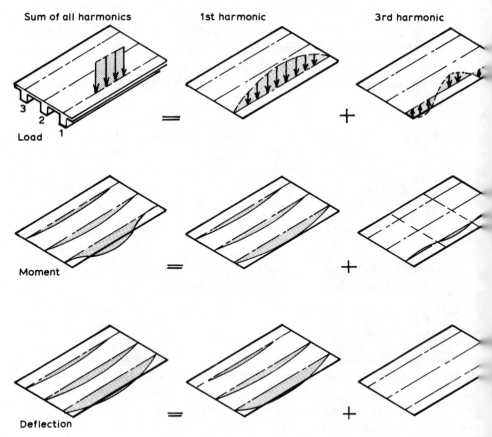

Fig. 12.7 Distribution of load, moment and deflection across a bridge deck.

beams 2 and 3 by the vertical shear associated with out-of-plane flexure and torsion of the slab and some by the in-plane shear of the slab. Although the slab is thin by comparison with the depth of the beam, its span between beams is short, giving a stiffness (for spanning between beams) comparable to that of the beams (spanning between abutments). Thus when beam 1 deflects, a significant proportion of the first harmonic of moment is transferred through the slab to beam 2. This is shown in Fig. 12.7, column 2, in which it can be seen that beam 1 is carrying about 60 per cent of the first harmonic bending moment while beams 2 and 3 are carrying about 25 and 15 per cent respectively. Precise amounts depend on the structure's form and dimensions. It should also be noted that, for a deck composed of identical beams, the distribution of first harmonics is identical for deflections and bending moments because for any harmonic component, bending moment is proportional to deflection.

12.3.2 Concentration of high harmonics

Under the action of the third harmonic of loading, the bridge deck in Fig. 12.7 is, in effect, simply supported between points of contraflexure of the third harmonic sinewave. Its span is effectively one third of the first harmonic span and, consequently, its longitudinal bending stiffness becomes 27 times that of the first harmonic. On the other hand, transverse bending of the slab is as before, with the slab spanning between beams. Consequently, when the loaded beam deflects under the action of the third harmonic of the load, nearly all the load is carried longitudinally by the very stiff beam, and very little transverse distribution takes place. A deck distributing the first harmonic in the manner of Fig. 12.7, column 2, would retain approximately 90 per cent of the third harmonic in the loaded beam, distributing only 10 per cent to the adjacent beam as shown in column 3.

As a working hypothesis, it is often convenient to assume that the first harmonic of the load is distributed transversely as shown in Fig. 12.7, column 2 while higher harmonics remain concentrated in the loaded beam. This is strictly only applicable to simply supported bridge decks with 'beams', as defined in Chapter 10, at centres greater than about one tenth of the span. However, it is often possible to analyse midspan sections of continuous bridge decks by considering the parts of the deck between points of contraflexure as being simply supported. In the general analysis of bridge decks which are continuous or have beams at close centres, it is necessary to analyse the higher harmonics, up to the level where distribution is considered negligible.

Another way of looking at the structural property of non-distribution of higher harmonics is to consider it as an example of Saint-Venant's Principle.

212 Bridge Deck Behaviour

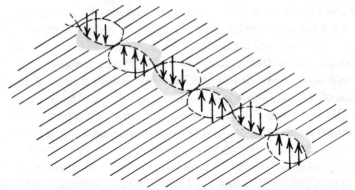

Fig. 12.8 Localized effect of high harmonic load.

Fig. 12.8 shows part of a slab subject to a high harmonic load. In the region of the wave shown, the slab must deflect downwards under the downward half wave and upwards along the upward half wave. However, at a point further than a half wavelength away, the effects of the downward and upward loads virtually cancel out. In other words, local variation in the load only affects a width similar to the length over which the variation occurs, while over a larger area the large number of upward and downward forces are self-cancelling. Consequently, if an abnormal vehicle is standing on a bridge deck, precise details of distribution of the point loads has little effect on the magnitude of bending in the beams, whose scale is much larger than distances between the wheels. But these distances do affect the local slab bending moments because the span of the slab is of the same scale. For this reason it is valid to superpose separate analyses for overall distributed bending moments and for local slab moments under the wheels. The latter can be obtained from the equations of Westergaard [2, 4] or the influence charts of Pucher [3]. However, it is important that the correct loads be applied in the analysis of overall deck moments, and that correct boundary conditions are assumed in the local bending analysis.

When a bridge deck is stiffened longitudinally by a few beams or webs at wide centres, it is important that the load should not simply be statically distributed between beams. The transverse flexural behaviour of the slab can be compared to that of a continuous beam with rotationally stiff elastic supports. In the load distribution analysis, the true load can be replaced by the fixed end moments and shear forces, and it should not be replaced by statically distributed loads on the supports.

Strictly, the fixed edge moments and shear forces should be obtained from the local load analysis using Pucher [3] with assumed fixed edge boundary conditions.

In contrast, loads can usually be statically redistributed locally along beams since there are no sudden variations in stiffness in that direction. Similarly, local statical redistribution of loads on a thick slab deck has little effect on the distributed load behaviour.

12.3.3 Example of separation of effects of low and high harmonics

Fig. 12.9 shows how the distribution of shear force near an abutment can be found by superposing the distributed first harmonic and the undistributed higher harmonics. Fig. 12.9a shows the point loads of two wheels of a vehicle near the abutment. (b) shows the shear force diagram for the bridge as a whole. The total shear force, i.e. sum of all harmonics, which is shown shaded was obtained using simple beam theory. The first harmonic component (shown dotted) was found using equations of Section 12.2. As explained previously, the first harmonic distributes significantly and this is shown in (c). The precise distribution relevant to the deck construction is obtained from charts of Chapter 10. The sum of the higher harmonics is the difference in (b) between the total shear force and the dotted first harmonic. These higher harmonics are assumed to remain undistributed and are applied solely to the loaded beam as in (d). The final shear

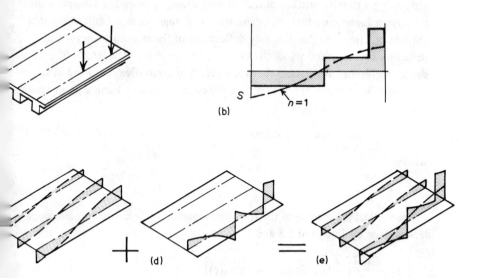

Fig. 12.9 Superposition of distributed first harmonic and undistributed higher harmonics of shear force. (a) Loads (b) total shear force diagram (c) distributed first harmonic (d) undistributed higher harmonics (e) final shear force distribution.

force distribution in (e) is found by recombining the distributed first harmonic in (c) with the undistributed higher harmonics in (d).

It is evident in Fig. 12.9e that the final distribution of shear force differs significantly from the distribution of the first harmonic (c) obtained from charts. This is because the higher harmonics form a significant part of the total shear force when the load is near one abutment. Appendix A Fig. A.3 gives first harmonic and higher harmonic components for several design loads. It can be seen that, in general, the first harmonic component of moment, slope or deflection approximates closely to the total function, and it is only for shear force near the load that the discrepancy can be very significant.

12.4 Harmonic analysis of plane decks

Several publications listed in the references [1, 4–7] outline the application of harmonic theory to the analysis of various types of deck in greater detail than is possible here. In particular for plane decks Rowe [4] outlines orthotropic plate theory and Hendry and Jaeger [5] describe the analysis of beam-and-slab decks and grid frameworks. To demonstrate here how harmonic analysis can be used, the artificially simplified deck of Fig. 12.10 is analysed below.

The beam-and-plank deck of Fig. 12.10a is constructed of three box beams supporting a running surface of transversely spanning planks. To simplify the problem it is assumed that the beams have such high torsional stiffnesses that they do not twist, so that the only deflections of the structure are vertical deflections w_1, w_2 and w_3 of the beams as shown in (b). The length of deck shown represents the distance between points of contraflexure of the nth harmonic. With origin at left end the deflections of beams 1 and 2 at any point are

$$w_1 \sin \alpha x \quad \text{and} \quad w_2 \sin \alpha x$$

where (12.9)

$$\alpha = \frac{n\pi}{L}.$$

The vertical shear force in the transverse spanning planks per unit length of deck is given by Equation 2.8 and is

$$s_{12} = \frac{6Ei}{l^2} \left[\frac{2}{l} (w_1 \sin \alpha x - w_2 \sin \alpha x) \right]$$

$$= \frac{12Ei}{l^3} (w_1 - w_2) \sin \alpha x \qquad (12.10)$$

where i = moment of inertia of planks per unit length of deck.

Harmonic Analysis and Folded Plate Theory 215

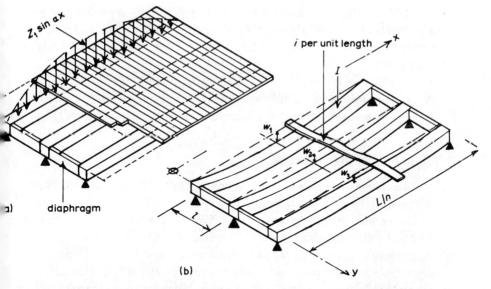

Fig. 12.10 (a) Loads and (b) deflections of beam-and-plank deck.

If $Z_1 \sin \alpha x$ is the nth harmonic of the applied load above beam 1, the net vertical load on beam 1 at any point is

$$Z_1 \sin \alpha x - s_{12} = \left[Z_1 - \frac{12Ei}{l^3}(w_1 - w_2)\right] \sin \alpha x. \quad (12.11)$$

It is shown in Equations 12.8 that if a beam of inertia I is subjected to load $b_n \sin \alpha x$, its deflection is $(b_n/EI\alpha^4) \sin \alpha x$. Consequently, if the load on beam 1 is given by Equation 12.11 and the deflection is $w_1 \sin \alpha x$, then

$$w_1 \sin \alpha x = \frac{1}{EI\alpha^4}\left[Z_1 - \frac{12Ei}{l^3}(w_1 - w_2)\right] \sin \alpha x$$

which can be written

$$\left(EI\alpha^4 + \frac{12Ei}{l^3}\right)w_1 - \frac{12Ei}{l^3} w_2 + 0w_3 = Z_1. \quad (12.12)$$

Similar equations can be obtained for beams 2 and 3 for equilibrium of bending load on beam, distributing load in planks, and applied load

$$-\frac{12Ei}{l^3}w_1 + \left(\frac{12Ei}{l^3} + EI\alpha^4 + \frac{12Ei}{l^3}\right)w_2 - \frac{12Ei}{l^3}w_3 = Z_2 \quad (12.13)$$

$$0w_1 - \frac{12Ei}{l^3}w_2 + \left(\frac{12Ei}{l^3} + EI\alpha^4\right)w_3 = Z_3 \quad (12.14)$$

Equations 12.12–12.14 can be solved to give values of w_1, w_2 and w_3 appropriate to the particular loads Z_1, Z_2 and Z_3. These deflections are then back substituted into Equations 12.10 and 12.8 to give the forces in the planks and beams.

The above procedure is followed for every harmonic of the load for which distribution is significant. Then by summing all the harmonics of beam forces, etc., the total distribution of load throughout the deck is obtained.

12.5 Folded plate analysis

In the example of Section 12.4, Equations 12.12–12.14 related the equilibrium of forces on the joint at each beam between the beam itself, the planks spanning to each or one side, and the applied load. The problem was simple because each joint had only one degree of freedom: vertical deflection. In contrast, each longitudinal joint of the folded plate structures in Fig. 12.1c, d and e has four degrees of freedom: vertical deflection, rotation about longitudinal axis, sideways deflection and warping displacement along the line of the joint. All other displacements of a point, such as rotation about the transverse axis, can be thought of as differentials of the four above. Though very much more complicated, the method of analysis is basically the same. The rigorous 'elastic' method of analysis of folded plate structures was derived by Goldberg and Leve [6] and presented in matrix form for computer analysis by DeFries-Skene and Scordelis [7]. Without the shorthand of matrix algebra and the numerical capacity of large computers, accurate solutions would not be practical. For example, with four degrees of freedom per joint, the number of simultaneous equations needed for the analysis of the deck of Fig. 12.1d would be at least 14 × 4 = 56. In practice it is often convenient to increase the number of plate strips by placing additional longitudinal joints between the slab/web intersections, and the number of simultaneous equations is likely to be closer to 120 for this example.

In addition to having many more systems of forces, the relationships between the forces and displacements in any one plate are much more complex than Equations 12.10 or 12.8. Figs. 12.11 and 12.12 show the edge forces and displacements on a plate strip separated into those causing out-of-plane flexure and twisting of the plate and those causing in-plane deformation. One of the basic assumptions of folded plate theory is that in-plane and out-of-plane behaviours are independent. A second basic assumption is that the end of every plate is restrained against out-of-plane displacement and rotation (w and ϕ in Fig. 12.11) and against in-plane lateral displacement (v in Fig. 12.12), but is free to warp (u in Fig. 12.12). These support restrictions ensure that the deck is simply supported and that the harmonic analysis has the simplified coefficients described in Section 12.2. Such restrictions on deflection are identical to assuming that at each end of the deck there is a right diaphragm which prevents all displacements within its plane.

The stiffness relationships between the harmonic amplitudes of the forces on one edge of a plate strip and the displacements of the two edges as shown in Figs. 12.11 and 12.12 can be written in matrix form:

$$\begin{bmatrix} r_{12} \\ p_{12} \\ s_{12} \\ m_{12} \end{bmatrix} = \begin{bmatrix} g_{11} & k_{11} & 0 & 0 \\ k_{11} & n_{11} & 0 & 0 \\ 0 & 0 & a_{11} & b_{11} \\ 0 & 0 & b_{11} & c_{11} \end{bmatrix} \begin{bmatrix} u_1 \\ v_1 \\ w_1 \\ \phi_1 \end{bmatrix}$$

$$+ \begin{bmatrix} g_{12} & k_{12} & 0 & 0 \\ -k_{12} & n_{12} & 0 & 0 \\ 0 & 0 & a_{12} & b_{12} \\ 0 & 0 & -b_{12} & c_{12} \end{bmatrix} \begin{bmatrix} u_2 \\ v_2 \\ w_2 \\ \phi_2 \end{bmatrix}$$

or $r_1 = k_{11} u_1 + k_{12} u_2$. (12.15)

Expressions for the coefficients are given in references [6 and 7] and in the rigorous 'elastic' method every coefficient is complicated. For example

$$g_{11} = \frac{Ed}{(1+v)^2} \frac{n\pi}{L} \left[-\frac{\sinh\left(\frac{n\pi l}{2L}\right)}{\frac{n\pi l}{2L} \operatorname{csch}\left(\frac{n\pi l}{2L}\right) - \frac{3-v}{1+v} \cosh\left(\frac{n\pi l}{2L}\right)} \right.$$

$$\left. + \frac{\cosh\left(\frac{n\pi l}{2L}\right)}{\frac{n\pi l}{2L} \operatorname{sech}\left(\frac{n\pi l}{2L}\right) + \frac{3-v}{1+v} \sinh\left(\frac{n\pi l}{2L}\right)} \right] \qquad (12.16)$$

where d is the thickness of the plate.

(A considerable simplification is introduced to the theory in what is called the 'ordinary' method which is the basis of some computer programs. In this method the plates are treated as transversely spanning strips for out-of-plane flexure, and as simple beams for in-plane deformation. While for some roof

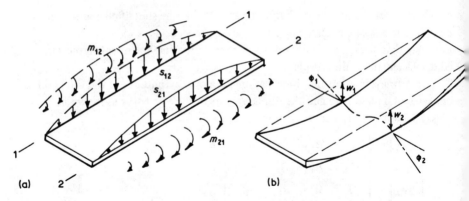

Fig. 12.11   (a) Out-of-plane loads and (b) deflections of plate.

structures little error is introduced, for bridges the neglect of in-plane shear deformation of slabs can introduce large errors.)

Before equilibrium equations can be written for the forces acting on a joint, it is necessary to transform the edge forces and displacement on a plate from the local coordinate system of the plate in Fig. 12.13a to the global coordinate system of the structure in (b) where the bar indicates a global variable. Resolving the local forces of (a) in the global directions we obtain

$$\begin{bmatrix} \bar{r} \\ \bar{p} \\ \bar{s} \\ \bar{m} \end{bmatrix} = \begin{bmatrix} 1 & 0 & 0 & 0 \\ 0 & \cos\theta & -\sin\theta & 0 \\ 0 & \sin\theta & \cos\theta & 0 \\ 0 & 0 & 0 & 1 \end{bmatrix} \begin{bmatrix} r \\ p \\ s \\ m \end{bmatrix}$$

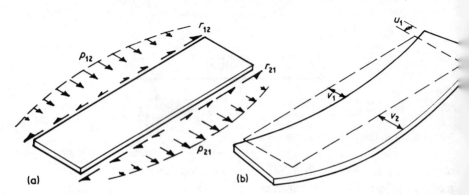

Fig. 12.12   (a) In-plane loads and (b) deflections of plate.

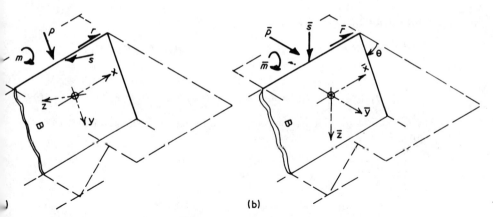

Fig. 12.13 (a) Local and (b) global forces on edge of plate.

or
$$\bar{r} = t\, r. \tag{12.17}$$

In addition, compatability of local displacements $u\ v\ w\ \phi$ and global displacements $\bar{u}\ \bar{v}\ \bar{w}\ \bar{\phi}$ requires

$$\begin{bmatrix} u \\ v \\ w \\ \phi \end{bmatrix} = \begin{bmatrix} 1 & 0 & 0 & 0 \\ 0 & \cos\theta & \sin\theta & 0 \\ 0 & -\sin\theta & \cos\theta & 0 \\ 0 & 0 & 0 & 1 \end{bmatrix} \begin{bmatrix} \bar{u} \\ \bar{v} \\ \bar{w} \\ \bar{\phi} \end{bmatrix}$$

or
$$u = t^t \tag{12.18}$$

where t^t is the transpose of t.

From Equations 12.16–12.18 we can obtain the stiffness relationships between global edge forces and displacements of a plate

$$\begin{aligned} \bar{r}_1 &= t k_{11} t^t \bar{u}_1 + t k_{12} t^t \bar{u}_2 \\ &= \bar{k}_{11} \bar{u}_1 + \bar{k}_{12} \bar{u}_2. \end{aligned} \tag{12.19}$$

The calculation of the transformation $t\, k\, t^t$ is a relatively simple procedure with modern computers.

Once the stiffness coefficients of all the plates have been calculated in terms of global forces and displacements, all the forces on each joint can be summed to give equilibrium equations such as for joint 2 in Fig. 12.14:

$$\bar{r}_{21} + \bar{r}_{23} + \bar{r}_{24} = \bar{x}_2 \tag{12.20}$$

which using Equation 12.19 gives

$$\bar{k}_{A_{21}}\bar{u}_1 + \bar{k}_{A_{22}}\bar{u}_2 + \bar{k}_{B_{22}}\bar{u}_2 + \bar{k}_{B_{23}}\bar{u}_3 + \bar{k}_{C_{22}}\bar{u}_2 + \bar{k}_{C_{24}}\bar{u}_4 = x_2$$

or

$$\bar{k}_{A_{21}}\bar{u}_1 + (\bar{k}_{A_{22}} + \bar{k}_{B_{22}} + \bar{k}_{C_{22}})\bar{u}_2 + \bar{k}_{B_{23}}\bar{u}_3 + \bar{k}_{C_{24}}\bar{u}_4 = \bar{x}_2. \qquad (12.21)$$

When Equation 12.21 is written with the equivalent equations of joints 1, 3, 4, 5 and 6 we obtain

$$\begin{bmatrix} \bar{k}_{A_{11}} & \bar{k}_{A_{12}} & 0 & 0 & 0 & 0 \\ \bar{k}_{A_{21}} & (\bar{k}_{A_{22}}+\bar{k}_{B_{22}}+\bar{k}_{C_{22}}) & \bar{k}_{B_{23}} & \bar{k}_{C_{24}} & 0 & 0 \\ 0 & \bar{k}_{B_{32}} & (\bar{k}_{B_{33}}+\bar{k}_{D_{33}}) & 0 & \bar{k}_{D_{35}} & 0 \\ 0 & \bar{k}_{C_{42}} & 0 & (\bar{k}_{C_{44}}+\bar{k}_{E_{44}}+\bar{k}_{F_{44}}) & \bar{k}_{E_{45}} & \bar{k}_{F_{46}} \\ 0 & 0 & \bar{k}_{D_{53}} & \bar{k}_{E_{54}} & (\bar{k}_{D_{55}}+\bar{k}_{E_{55}}) & 0 \\ 0 & 0 & 0 & \bar{k}_{E_{64}} & 0 & \bar{k}_{F_{66}} \end{bmatrix} \begin{bmatrix} \bar{u}_1 \\ \bar{u}_2 \\ \bar{u}_3 \\ \bar{u}_4 \\ \bar{u}_5 \\ \bar{u}_6 \end{bmatrix}$$

$$= \begin{bmatrix} \bar{x}_1 \\ \bar{x}_2 \\ \bar{x}_3 \\ \bar{x}_4 \\ \bar{x}_5 \\ \bar{x}_6 \end{bmatrix}$$

or

$$\bar{K}\bar{u} = \bar{X} \qquad (12.22)$$

when $\bar{K}$, $\bar{u}$ and $\bar{X}$ are the stiffness, displacement and load matrices, respectively, for the whole structure. Matrix Equations 12.22 represents 24 simultaneous equations for the sets of four displacements at all six joints. Solution of the equations gives the values of the joint displacements appropriate to the particular load. Subsequent back substitutions of these displacements into the stiffness equations of the individual plates permit calculation of the forces and stresses in the plates. As for the example of Section 12.4, the above procedure must be repeated for every harmonic of load for which the distribution is significant, and the results added.

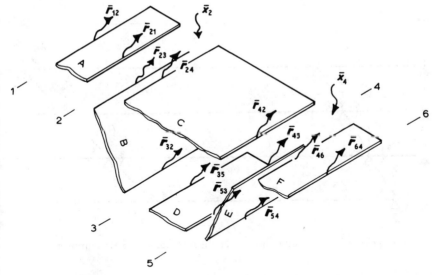

Fig. 12.14 Forces on edges of plates of structure.

12.6 Continuous and skew decks

In Section 12.2 it is assumed that the bridge deck is right and simply supported. Then by choosing only harmonics of load which have points of contraflexure coincident with the supports, the bending moment and deflection automatically become zero at the supports. If on the other hand the deck has spring supports, internal supports or skew abutments, the analysis becomes very much more cumbersome. If the deck has right abutments but is continuous over an internal support such as J in Fig. 12.15, the deck can still be analysed as if it is simply supported between abutments. First the distributions of moment, deflection, etc. are found independently for a unit reaction $R_J = 1$ and for the live load. Then the distribution due to R_J is scaled up or down so that its deflection at J cancels that due to the live load (or for a spring support leaves a residual deflection equal to $R_J \times$ spring stiffness). The distributions are superposed to give the distribution for the continuous deck with no deflection at J. The same principle can be employed for a deck with any support conditions including skew abutments ABCD, EFGH in Fig. 12.15. The skew deck can be thought of as part of a longer structure of arbitrary length L. If the deck has the simple structure of Fig. 12.10, the reactant forces and moments R_A, M_A, etc., are applied to the structure so that the combined distribution of these loads and the live load give no net deflection or moment at each skew support point. More

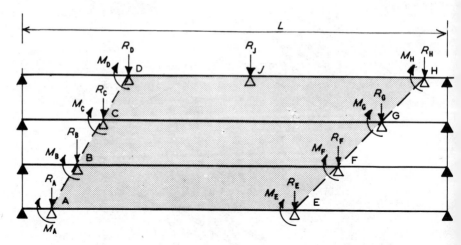

Fig. 12.15 Redundant forces at internal and skew supports.

complex structures such as Fig. 12.14 are solved by introducing as many reactant forces at the supports as there are independent deflections or forces which must be eliminated or balanced.

12.7 Errors of harmonics near discontinuities

Computer programs employing harmonic analysis can only consider a limited number of harmonics (say ten or one hundred) and ignore the higher harmonics. If the sum of the considered harmonics of load or shear force, etc., is plotted in the region of a discontinuity as in Fig. 12.16, the line violently oscillates close to the discontinuity. Increasing the number of considered harmonics moves the oscillation closer to the discontinuity, but the amplitude is not reduced. At the limit the oscillation still exists but is infinitely narrow. This characteristic of harmonic analysis, called Gibb's Phenomenon, can lead to significant errors in the output from folded plate and finite strip computer programs in the region to

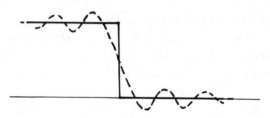

Fig. 12.16 Gibb's phenomenon.

each side of a discontinuity within two or three wavelengths of the highest harmonic considered. Consequently, unless the computer plots out the oscillating distribution along the deck, results output for discontinuous functions such as shear force over a pier should be considered suspect in the region of the discontinuity. These errors can largely be avoided if the analysis lumps all the undistributed higher harmonics in the loaded member as in Section 12.3.3.

REFERENCES

1. Kreyszig, E. (1962), *Advanced Engineering Mathematics,* John Wiley, New York.
2. Westergaard, H. M. (1930), 'Computations of stresses in bridge slabs due to wheel loads,' *Public Roads,* 2, 1–23.
3. Pucher, A. (1964), *Influence Surfaces of Elastic Plates,* Springer Verlag, Wien and New York.
4. Rowe, R. E. (1962), *Concrete Bridge Design,* C. R. Books Ltd, London.
5. Hendry, A. W. and Jaeger, L. G. (1958), *The Analysis of Grid Frameworks and Related Structures,* Chatto and Windus, London.
6. Goldberg, J. E. and Leve, H. L. (1957). 'Theory of prismatic folded plate structures,' International Association for Bridge and Structural Engineering, Zurich, No 87 pp. 71–72.
7. DeFries-Skene, A. and Scordelis, A. C. (1964), 'Direct stiffness solution for folded plates,' *Proc. ASCE,* ST 4, pp. 15–47.

13
Finite element method

13.1 Introduction

The finite element method is a technique for analysing complicated structures by notionally cutting up the continuum of the prototype into a number of small elements which are connected at discrete joints called nodes. For each element, approximate stiffness equations are derived relating the displacements of the nodes to the node forces between elements and, in the same way that slope-deflection equations can be solved for joints in a continuous beam, an electronic computer is used to solve the very large number of simultaneous equations that relate node forces and displacements. Since the basic principle of subdivision of the structure into simple elements can be applied to structures of all forms and complexity, there is no logical limit to the type of structure that can be analysed if the computer program is written in the appropriate form. Consequently, finite elements provide the most versatile method of analysis available at present, and for some structures the only practical method. However, the quantity of computation can be enormous and expensive so that often the cost cannot be justified for run-of-the-mill structures. Furthermore, the numerous different theoretical formulations of element stiffness characteristics all require approximations which in different ways affect the accuracy and applicability of the method. Further research and development is required before the method will have the ease of use and reliability of the simpler methods of bridge deck analysis described in previous chapters.

The technique was pioneered for two-dimensional elastic structures by Turner *et al.* [1] and Clough [2] during the 1950s. Since then a very considerable development has been made by many people. This chapter does little more than demonstrate the basic physical principles. Much more detailed and comprehensive descriptions of the method are given in the books of Zienkiewicz and Cheung [3, 4], Holand and Bell [5] and Desai and Abel [6], with useful demonstrations of bridge analyses in Rockey, Bannister and Evans [7].

13.2 Two-dimensional plane stress elements

The finite element method is first demonstrated in relation to the analysis of the plane stress (or 'in-plane' or 'membrane') behaviour of flat plates. This is one of the simplest applications of the method, ignoring continuous beams, and it is relevant to the in-plane actions of the slabs and webs of beam-and-slab and cellular bridge decks. Out-of-plane bending is more complicated and is discussed in Section 13.3.

Fig. 13.1a shows an elevation of a beam which in (b) is subjected to pure bending. This simple structure is chosen for the example because its behaviour is well known to civil engineers, but the general principles of the following discussion can be applied to a plate of any shape subjected to any system of in-plane forces. For analysis, the structure is considered as in (c) to be made up of a large number of triangular elements connected together only at the corners. The triangular elements are drawn separated except at nodes to emphasize that there is no force interaction along the cuts. When this articulated structure is subjected to pure bending as in (d), the deformation of each element only depends on the movements of the nodes. In the simplest of element models it is assumed that the strains within each element are uniform during distortion with the triangle edges remaining straight, as in Fig. 13.2a. Thus, while the 45° lines in the prototype of Fig. 13.1b deflect in curves, those in the model of Fig. 13.1d deflect as strings of short straight lines. The difference between the model and prototype can be reduced if the model is composed of a larger number of smaller sized elements. At the theoretical limit with an infinite number of infinitely small elements, the model is effectively a continuum like the prototype.

The stiffness equation of this simple element model can be derived directly from the theory of elasticity plane stress equations. If the triangle of Fig. 13.2a is assumed to distort with internal straight lines and edges remaining straight, the

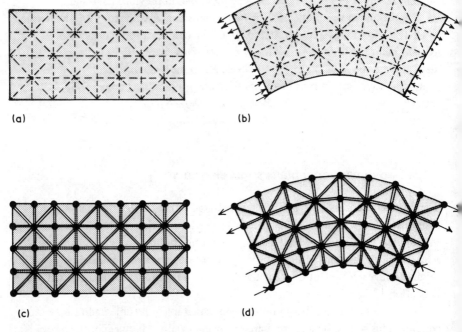

Fig. 13.1 Bending of beam. (a and b) Prototype (c and d) finite element model.

displacement field is linear and can be expressed as

$$u = \alpha_1 + \alpha_2 x + \alpha_3 y$$
$$v = \alpha_4 + \alpha_5 x + \alpha_6 y. \quad (13.1)$$

The strains are given by

$$\epsilon_x = \frac{\partial u}{\partial x} = \alpha_2$$

$$\epsilon_y = \frac{\partial v}{\partial y} = \alpha_6$$

$$\gamma_{xy} = \frac{\partial u}{\partial y} + \frac{\partial v}{\partial x} = \alpha_3 + \alpha_5. \quad (13.2)$$

If the three nodes have coordinates (x_1, y_1), etc. and displace u_1, v_1, etc. then Equations 13.1 can be written for the six displacements etc. For example the three equations for u are

$$u_1 = \alpha_1 + \alpha_2 x_1 + \alpha_3 y_1$$
$$u_2 = \alpha_1 + \alpha_2 x_2 + \alpha_3 y_2 \qquad (13.3)$$
$$u_3 = \alpha_1 + \alpha_2 x_3 + \alpha_3 y_3$$

which can be solved to give α_1, α_2 and α_3, and with α_4, α_5 and α_6, from the equations for v we obtain

$$\epsilon_x = \alpha_2 = \frac{(u_1 - u_2)(y_2 - y_3) - (u_2 - u_3)(y_1 - y_2)}{(x_1 - x_2)(y_2 - y_3) - (x_2 - x_3)(y_1 - y_2)}$$

$$\epsilon_y = \alpha_6 = \frac{(v_1 - v_2)(x_2 - x_3) - (v_2 - v_3)(x_1 - x_2)}{(x_1 - x_2)(y_2 - y_3) - (x_2 - x_3)(y_1 - y_2)} \qquad (13.4)$$

$$\gamma_{xy} = \alpha_3 + \alpha_5 = -\frac{(u_1 - u_2)(x_2 - x_3) - (u_2 - u_3)(x_1 - x_2)}{(x_1 - x_2)(y_2 - y_3) - (x_2 - x_3)(y_1 - y_2)}$$
$$+ \frac{(v_1 - v_2)(y_2 - y_3) - (v_2 - v_3)(y_1 - y_2)}{(x_1 - x_2)(y_2 - y_3) - (x_2 - x_3)(y_1 - y_2)}$$

so relating element strains to node displacements.

The stresses in the element can be found from the elastic theory equations for plane stress

$$\sigma_x = \frac{E}{(1 - \nu^2)} \epsilon_x + \frac{\nu E}{(1 - \nu^2)} \epsilon_y$$

$$\sigma_y = \frac{\nu E}{(1 - \nu^2)} \epsilon_x + \frac{E}{(1 - \nu^2)} \epsilon_y \qquad (13.5)$$

$$\tau_{xy} = \frac{E}{2(1 + \nu)} \gamma_{xy}.$$

If the element were part of a continuum, stresses of this magnitude would cross the boundaries as shown in Fig. 13.2b. However, these stresses are represented by the node forces of (c) with the stresses on each edge statically distributed to the neighbouring nodes. Since the stresses forming U_1 and V_1 are in equilibrium with the stresses (σ_x, σ_y, τ_{xy}) on the internal cuts shown through the mid points

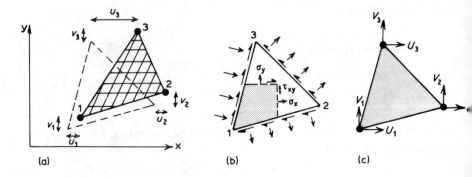

Fig. 13.2 Node displacements and forces on simple plane stress triangular element. (a) Displacements (b) stresses (c) node forces.

of the sides we can resolve

$$U_1 = \sigma_x \frac{(y_2 - y_3)}{2} - \tau_{xy} \frac{(x_2 - x_3)}{2}$$

$$V_1 = -\sigma_y \frac{(x_2 - x_3)}{2} + \tau_{xy} \frac{(y_2 - y_3)}{2}$$

(13.6)

substituting for σ_x, etc. from Equation 13.5,

$$U_1 = \frac{E}{(1 - \nu^2)} \frac{(y_2 - y_3)}{2} \epsilon_x + \frac{\nu E}{(1 - \nu^2)} \frac{(y_2 - y_3)}{2} \epsilon_y$$

$$- \frac{E}{2(1 + \nu)} \frac{(x_2 - x_3)}{2} \gamma_{xy}$$

$$V_1 = -\frac{\nu E}{(1 - \nu^2)} \frac{(x_2 - x_3)}{2} \epsilon_x - \frac{E}{(1 - \nu^2)} \frac{(x_2 - x_3)}{2} \epsilon_y$$

$$+ \frac{E}{2(1 + \nu)} \frac{(y_2 - y_3)}{2} \gamma_{xy}.$$

(13.7)

These equations can be combined with Equation 13.4 to give stiffness equations for node 1 as a point of the element

$$U_1 = k_1 u_1$$

(13.8)

relating forces on node 1 to displacements of nodes 1, 2 and 3. Then consideration of the equilibrium of node 1 under the sum of forces from all elements adjoining node 1, together with applied loads X_1, provides stiffness equations for node 1, as a point in the structure

$$X_1 = K_1 \text{ [displacements of node 1 and neighbours]}. \tag{13.9}$$

Such equations can be derived for every node in the model structure, forming in all $2N$ simultaneous equations for N nodes. These equations are solved by the computer to give the displacements (u, v) at every node. From these the strains and stresses in every element can be calculated using Equations 13.4 and 13.5.

An example of the stress distribution computed from a coarse mesh model subjected to pure bending is shown in Fig. 13.3a. The associated distortion of the elements is shown in (b). Since the stresses are only induced here in the x-direction, the stress in each element is simply proportional to the shortening of the edge parallel to the x-axis. Thus it can be seen in (b) that the x-compressions of elements b and c are the same and equal to half that of a, while d is unstrained. Fig. 13.4a shows how the stresses in elements a, b, c and d form node forces. The forces are zero where the stresses on the edges on the two sides of the node cancel out. Fig. 13.4b shows the stress distributions that could be plotted for sections X-X and Y-Y in Fig. 13.3a. These can be compared with the exact stress distribution. It is evident that the stresses computed for individual elements are here misleading. However the average stresses round each node provide an estimate within 13 per cent of the exact figure.

For the more complex elements, it is not usually possible to relate node forces directly to element stresses and the assumed displacement field.

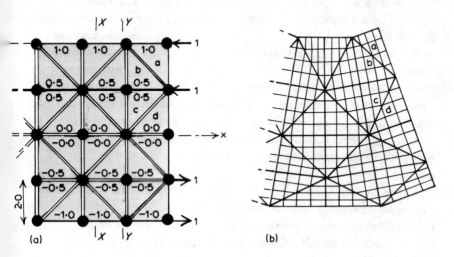

Fig. 13.3 Pure bending of triangular mesh. (a) Distribution of σ_x (b) displacement field.

230 *Bridge Deck Behaviour*

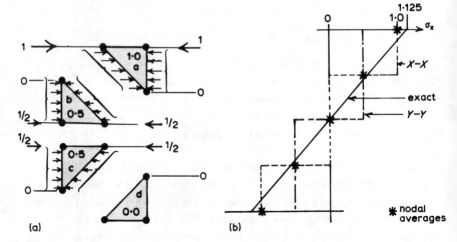

Fig. 13.4   (a) Stresses and node forces on elements   (b) stress distributions on sections.

The element stiffnesses are frequently derived from a consideration of the potential energy stored by the assumed displacement field. For each node force, the external work done during a virtual displacement is equated to the minimum increase in potential energy that can be stored by the displacement field. While this mathematical concept enables more complicated problems to be analysed with computing economy, it does mean that element stresses output by a program are not directly related by equilibrium to the applied loads. If the element displacement function is not appropriate to the problem, the output element stresses can be as low as 50 per cent or less of those necessary for equilibrium with the applied loads. Calculating the nodal averages does not necessarily make much difference. Although continual improvements in programs are reducing the possibility of such errors, it is advisable to make a hand check of the equilibrium of output element stresses and the applied loads wherever possible. Because significant discrepancies have been found in the past, many design engineers still place more confidence in a space frame analysis in which output member forces are automatically in equilibrium with applied loads.

A finer mesh is shown in Fig. 13.5 and it is evident that the nodal average stresses are now within 3 per cent of the exact figures in Fig. 13.4. In general, the finer the finite element mesh is, the more accurate are the results. A very coarse mesh as shown in Fig. 13.6 can be extremely inaccurate. Furthermore, the curvature of the beam, proportional to the difference between the top and bottom element stresses, is very much less than that for Fig. 13.5. This implies

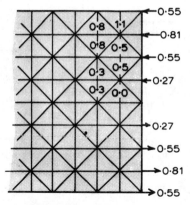

Fig. 13.5 More accurate stresses obtained with finer mesh.

that the coarse mesh is very much stiffer than the fine mesh. It might appear foolish even to consider a mesh as coarse as Fig. 13.6a. However, the example is a warning of the very large errors that might be introduced if this beam were a web of a box girder deck such as in (b) where at a first glance the mesh might not appear coarse. The problem highlights the importance of choosing the most appropriate element arrangement and of checking the solution when possible against a different type of analysis.

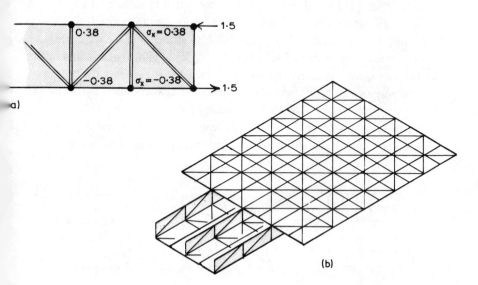

Fig. 13.6 (a) Inaccurate stresses from very coarse mesh (b) coarse mesh of box girder.

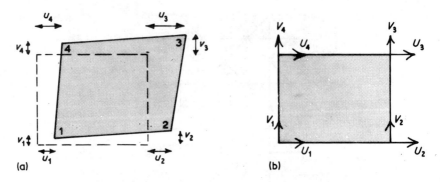

Fig. 13.7 Node displacements and forces on quadrilateral element.

The finite elements of a model do not have to be triangular; numerous other shapes including quadrilaterals, rectangles, parallelograms and other polygons are used. Fig. 13.7 shows the node displacements and forces of a rectangular element. The displacements can be assumed to have functions

$$u = \alpha_1 + \alpha_2 x + \alpha_3 y + \alpha_4 xy$$
$$v = \alpha_5 + \alpha_6 x + \alpha_7 y + \alpha_8 xy. \qquad (13.10)$$

The distortion is then similar to Fig. 13.8a with linear displacements (i.e. constant strain) along each edge. It can be noted that Equations 13.10 have eight unknowns which are found by solving the equations for the eight displacements $u_1, v_1, u_2, \ldots v_4$. An alternative approach is to assume that the rectangle has the stiffness characteristics of the pairs of triangles of either (b) or (c) or of the average of (b) and (c). Other rectangular elements have also been derived for various purposes using different displacement functions. Some in particular [7] have been evolved to represent webs bending in-plane.

In the preceding discussion it has been assumed that the model is made up of compatible elements, i.e. displacement functions are assumed so that adjacent

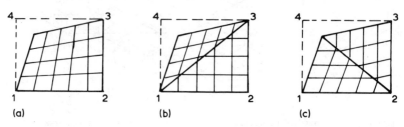

Fig. 13.8 Displacement fields in various simple quadrilateral elements.

points on elements on each side of a cut always remain adjacent. The stresses are discontinuous at the cuts, but the stress resultants at nodes are in equilibrium. By making the model distort in a specified way, it has less freedom and is stiffer than the prototype and computed stresses are lower bounds, always being less than exact solutions. This was demonstrated in Fig. 13.4. An alternative approach is to assume stress functions for the elements so that stresses are continuous across the edges of elements. Displacements are then discontinuous with relative displacements between edges of elements except at the nodes. The resulting model is then more flexible than the prototype and computed stresses are upper bounds, being greater than the exact ones. This flexibility method is not used very often because of its greater theoretical complexity.

13.3 Plate bending elements

The finite element technique is often used for analysis of plate bending behaviour of slab bridges. The basic concept and process is similar to that for plane stress. However the deflection and force variables are different, and the theoretical derivation of the element stiffness equations presents greater theoretical problems.

Fig. 13.9a shows a prototype slab deck bending under vertical loading, while (b) shows a possible system of subdivision of the model into a large number of triangular elements. The piece of prototype represented by an element experiences vertical deflection w (and curvatures which are differentials of w) as shown in Fig. 13.10a and is subjected to moments, torsions and vertical shear forces along its edges, as in (b). These edge forces are represented in the stiffness equations by the node forces shown in (d):

F_w = a vertical shear force
$F_{\theta x}$ = a moment about axis Ox
$F_{\theta y}$ = a moment about axis Oy.

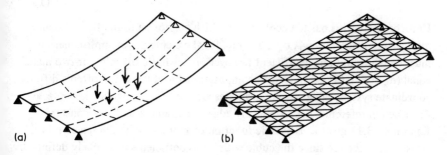

Fig. 13.9   Finite element model for plate bending of slab deck.

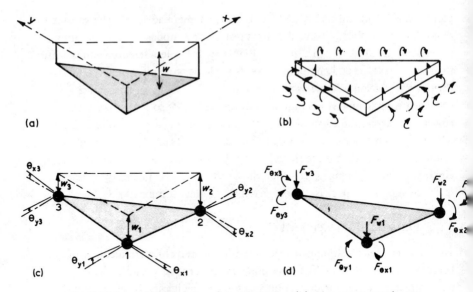

Fig. 13.10 (a) Triangular plate element (b) edge stress resultants (c) node displacements (d) node forces.

The node deflections in (c) appropriate to the forces in (d) are

w = vertical deflection
θ_x = rotation of node about Ox
θ_y = rotation of node about Oy.

To determine the stiffness of such an element, a displacement field must be assumed as was done for plane stress elements in Equation 13.1. A first choice might be the complete third order polynomial:

$$w = \alpha_1 + \alpha_2 x + \alpha_3 y + \alpha_4 x^2 + \alpha_5 xy + \alpha_6 y^2 + \alpha_7 x^3 + \alpha_8 x^2 y + \alpha_9 xy^2 + \alpha_{10} y$$

(13.11)

Unfortunately, this has ten coefficients which cannot be found from the nine displacement variables $w_1, \theta_{x_1}, \theta_{y_1}$, etc. for the element. An approximation must be made and one coefficient removed somewhat arbitrarily, or two made equal ($\alpha_8 = \alpha_9$). (This can lead to computational problems, and often a different coordinate system called area coordinates are used for triangular elements, as described in reference [3].) Along any edge, for example with y = constant, Equation 13.11 gives w as a cubic function of x; this will be the same on both sides of an interface since the cubic with four coefficients is uniquely defined by four node displacements. However the normal slope $\partial w/\partial y$ is a quadratic of x

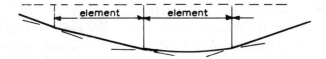

Fig. 13.11 Discontinuity of slope at element edges between nodes.

with three coefficients, and since there are only two defining node variables

$$\theta_{x_1} = \left(\frac{\partial w}{\partial y}\right)_1 \quad \text{and} \quad \theta_{x_2} = \left(\frac{\partial w}{\partial y}\right)_2$$

on the edge, the normal slope is not uniquely defined and can be different for the elements on two sides of an interface. Consequently, as is shown in Fig. 13.11, except at nodes, there is a kink at the interface between elements even though the vertical deflections are continuous. Thus, because the complete polynomial is too general for the number of defining degrees of freedom, the elements are not truly compatible, and are said to be 'non-conforming'.

A simple rectangular element such as Fig. 13.12a is also non-conforming, since the lowest order polynomial with sufficient coefficients to describe the deflections is fourth order with sixteen coefficients while the element has only twelve defining degrees of freedom. Numerous other bending elements have been proposed which are also non-conforming. However there are some, such as in Fig. 13.12b, which are made 'conforming'. In this example, the number of defining degrees of freedom of the element is increased by adding additional nodes at the mid points of each edge. At these nodes the normal slope ($\partial n/\partial s$) is the displacement variable, with bending moment the node force. Now a complete

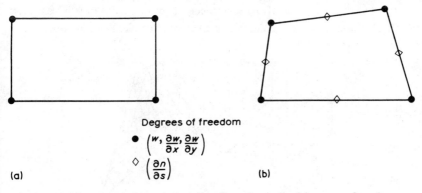

Fig. 13.12 Quadrilateral plate bending elements. (a) nonconforming (b) conforming.

fourth order polynomial with sixteen coefficients can be solved using the sixteen degrees of freedom. The normal slope which varies as a quadratic along any edge is uniquely defined by three values at the two end nodes and one middle node.

The stiffness equations for the elements are usually derived, as mentioned in Section 13.2, by a consideration of virtual work. The node is given a virtual displacement for each of the degrees of freedom, and the corresponding node forces are found by equating the externally applied work to the minimised increase in potential energy stored by the assumed displacement function.

Solutions obtained with compatible conforming elements are always lower bounds since the model structure is stiffer when the ways it can deform are completely specified. In contrast, solutions obtained with non-conforming elements are neither lower bounds nor upper bounds. They do not satisfy either of the following assumptions: compatability throughout the structure for a lower-bound minimum potential energy solution; or equilibrium throughout the structure for an upper-bound minimum complementary energy solution. A non-conforming element will give higher stresses than the lower bound obtained from a conforming element of the same shape because it has less stiffness due to its greater freedom in taking up a deflected shape. Sometimes, ironically, the simple non-conforming triangular or rectangular elements can give better solutions than much more sophisticated conforming elements. This is done at a very great savings in cost since the conforming elements are significantly more cumbersome for computer manipulation. In consequence, non-conforming triangular or quadrilateral elements are generally used for slab bridges. The triangle is usually the more popular because without difficulty it can be made to fit decks of complex plan shape as in Fig. 13.13.

Before leaving bending elements, it is worth mentioning the simple beam element which, as shown in Fig. 13.14, can be used to represent the stiffening due to a beam in the plane of the slab. Since the beam carries torsion as well as bending, its nodes have three degrees of freedom, just like the plate nodes, and

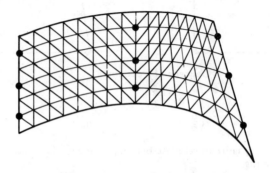

Fig. 13.13   Slab deck divided into triangular elements.

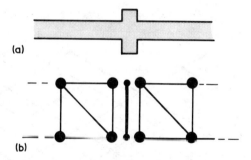

Fig. 13.14 Beam element located between triangular elements. (a) Section of slab stiffened by beam   (b) plan of finite elements with beam element.

the node forces are vertical shear, bending moment and torsion. It is the basic element of a grillage model, which itself can be thought of as a simple finite element model. The grillage is non-conforming since no consideration has been made for compatability of deflections or equilibrium of stresses between elements except at nodes. It should be noted that the in-plane beam of Fig. 13.14 is no better than a grillage beam at representing the slab-membrane action due to a beam located below the slab of a beam-and-slab deck. Consequently, it is unlikely that a two-dimensional plate bending finite element analysis will produce more reliable results than a grillage for such a deck. Only a three-dimensional analytical model could produce more accurate predictions. On the other hand, a two-dimensional finite element analysis with fine mesh should produce more accurate results than a grillage for a plane slab deck when Poisson's Ratio is large so that there is interaction in the plate bending Equations 3.5 or 3.9.

13.4 Three-dimensional plate structures and shell elements

A detailed investigation of a beam-and-slab or cellular bridge deck requires a three-dimensional analysis. Even though it is generally possible to approximate the behaviour of slabs and webs to thin plates, these must be arranged in a three-dimensional assemblage as in Fig. 13.15.

At every intersection of plates lying in different planes there is an interaction between the in-plane forces of one plate and the out-of-plane forces of the other, and vice versa. For this reason it is essential to use finite elements which can distort under plane stress as well as plate bending. Since it is assumed that for flat plates, in-plane and out-of-plane forces do not interact within the plate, the elements are in effect the same as a plane stress element (described in Section 13.2) in parallel with a plate bending element of Section 13.3.

It should be noted that even if the elements are conforming individually for

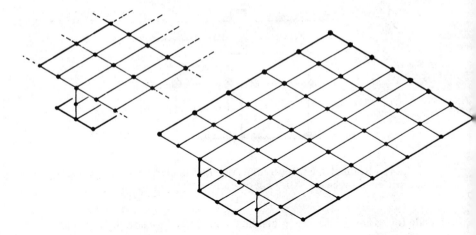

Fig. 13.15 Three-dimensional structures composed of plate elements.

plane stress and bending, the displacements are not usually compatible along the web/slab intersections except at nodes. For the reasons discussed in Section 13.2, the simple triangular element is not suitable for such an analysis unless an extremely fine mesh is used. However special elements are available which can represent the in-plane bending of the webs.

There is no logical limit to the cellular complexity, structural shape or support system of a bridge that can be analysed with a three-dimensional plate model. However every node must have six degrees of freedom, three deflections and three rotations, and solution of the very large number of stiffness equations generated for even relatively simple structures is expensive. Consequently, the method is usually only used to study the distribution of stress in one span or intricate part of a structure. These results are then used to interpret the distributions of stress resultants such as the overall moment that are output by the simpler models of continuous beam, grillage or space frame.

Shell structures such as arches can usually be analysed with plate elements as in Fig. 13.16a. However, shell elements as in (b) are available. In such elements the interaction of in-plane and out-of-plane forces takes place throughout the element and not just at the nodes.

13.5 Finite strips

Bridge decks which have the same cross-section from end to end can be analysed with a simple and economic type of finite element called a finite strip. The method, pioneered by Cheung [8], is very similar to folded plate analysis

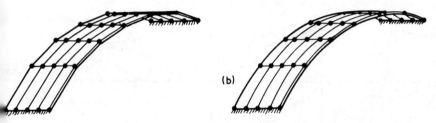

Fig. 13.16 Arched structures of (a) plate elements and (b) shell elements.

scribed in Chapter 12. The structure is assumed, as in Fig. 13.17a, to be made of finite elements called 'strips' which extend from one end of the deck to the other. The strips are connected by nodes which also run from one end to the other. Like folded plate theory, the displacement functions for in-plane and out-of-plane deformation of the strips are of the form

$$w, \theta, u \text{ or } v = \Sigma f(y) \sin\left(\frac{n\pi x}{L}\right) \tag{13.12}$$

where x is the direction along the structure and y is the direction across the strip. As explained in Chapter 12, harmonic analysis is greatly simplified if the deck is assumed to have right end supports with diaphragms to prevent displacement of the ends of the plates in the plane of the diaphragm. The

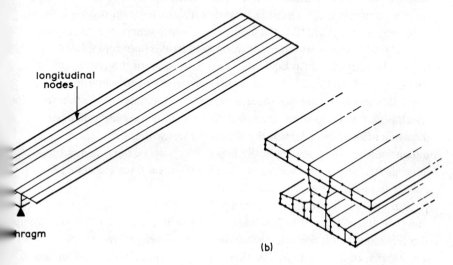

Fig. 13.17 (a) Finite strip model of box deck   (b) finite prism model.

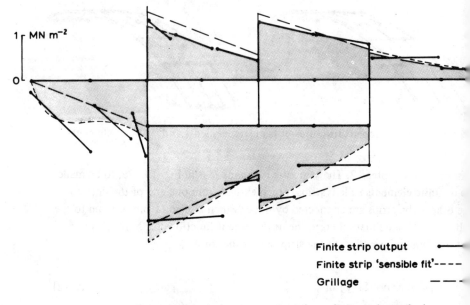

Fig. 13.18 Slab-in-plane shear stress output from finite strip analysis.

analytical procedure is also the same in that stiffness equations are obtained and solved for each harmonic component of the load in turn, and the results summed to give the total stress distribution. Furthermore, similar errors due to Gibb's Phenomenon can be encountered near discontinuities (see Section 12.7). In the finite strip method, the transverse functions $f(y)$ are assumed to be simple polynomials so that in effect the method is an approximation to the rigorous 'elastic' folded plate method in which these functions have complicated hyperbolic form similar to Equation 12.16. One result of this approximation is that calculated stresses are discontinuous transversely at strip interfaces. Fig. 13.18 shows the in-plane shear stress distribution output for a two-cell spine which is part of a two-spine concrete deck. The computer output indicates discontinuities of shear flow at the strip interfaces within slabs. These discontinuities, which are physically impossible, must be smoothed out to a 'sensible fit'. Also shown for comparison are the shear flows calculated from a grillage following the procedures of Chapter 5.

Harmonic analysis can also be used with two-dimensional displacement functions to analyse prismatic solid structures such as Fig. 13.17b. Furthermore this technique and finite strip analysis have been developed for curved circular structures with the harmonic function used for variations along circular arcs.

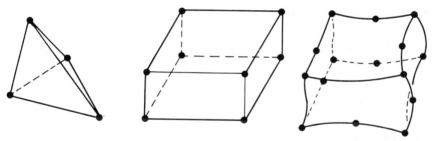

Fig. 13.19 Three-dimensional solid elements.

13.6 Three-dimensional elements

Three-dimensional solid elements are seldom used in the analysis of bridge decks because generally these structures are composed of the thinnest plates possible to minimize weight. Solid elements are used more often for the analysis of nuclear reactors and complex soil structures. The simplest elements consist of tetrahedra or hexahedra with nodes at the corners, as shown in Fig. 13.19a and b. If the mesh is fine, the nodes need only have three degrees of freedom for displacement in the three dimensions. More sophisticated elements have additional nodes in addition to those at the corners, with more degrees of freedom at each node.

13.7 Conclusion

The finite element method is the most powerful and versatile analytical method available at present because with a sufficiently large computer, the elastic behaviour of almost any structure can be analysed accurately. For this reason it is often requested by clients, or proposed to a client, to show that the most accurate analysis possible has been performed. Unfortunately, the method is cumbersome to use and is usually expensive. In addition, the choice of element type can be extremely critical and if incorrect, the results can be far more inaccurate than those predicted by simpler models such as grillage or space frame. However, perhaps the greatest drawback at present is that while the technique is developing so rapidly, the job of carrying out finite element computations is a full time occupation which cannot be carried out at the same time by the senior engineer responsible for the design. He is unlikely to have time to understand or verify the appropriateness of the element stiffnesses or to check the large quantity of computer data. This makes it difficult for him to place his confidence in the results, especially if the structure is too complicated for him to use simple physical reasoning to check orders of magnitude.

For these reasons, when commissioning a finite element analysis, it is advisable to check that the computer organization and employees involved have had plenty of experience of the program, and that the program has been well tested for similar structural problems. Furthermore, if the structure is too complicated to apply physical reasoning, it is worth commissioning an inexpensive simple space frame analysis as an independent check.

REFERENCES

1. Turner, M. J., Clough, R. W., Martin, H. C., and Topp, L. J. (1956), 'Stiffness and Deflection Analysis of Complex Structures', *J. Aero. Sci.*, **23**, 805–23.
2. Clough, R. W., (1960), 'The finite element in plane stress analysis,' Proc. 2nd A.S.C.E. Conf. on Electronic Computation, Pittsburgh, Pa., Sept.
3. Zienkiewicz, O. C. and Cheung, Y. K. (1967), *The Finite Element Method in Structural and Continuum Mechanics*, McGraw-Hill, New York and London.
4. Zienkiewicz, O. C. (1971), *The Finite Element Method in Engineering Science*, McGraw-Hill, London.
5. Holand, I. and Bell K. (1969), *Finite Element Method in Stress Analysis*, Tapier, Trondheim, Norway.
6. Desai, C. S. and Abel, J. F. (1972), *Introduction to the Finite Element Method*, Van Nostrand Reinhold, New York.
7. Rockey, K. C., Bannister, J. L. and Evans, H. R. (1971) (eds), *Developments in Bridge Design and Construction*, Crosby Lockwood, London.
8. Cheung, Y. K. (1968), 'The finite strip method in the analysis of elastic plates with two opposite simply supported ends,' *Proc. Inst. Civ. Eng.*, **40**, 1–7.

APPENDIX A

Product integrals. Functions of load on a single span. Harmonic components

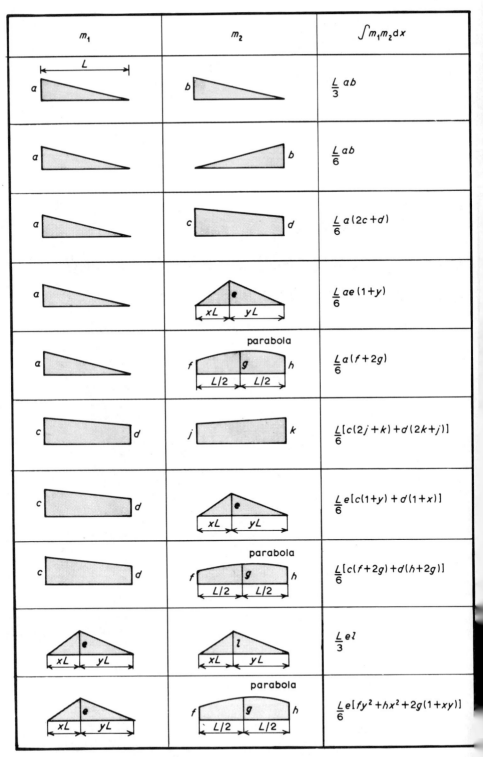

Fig. A.1 Product integrals.

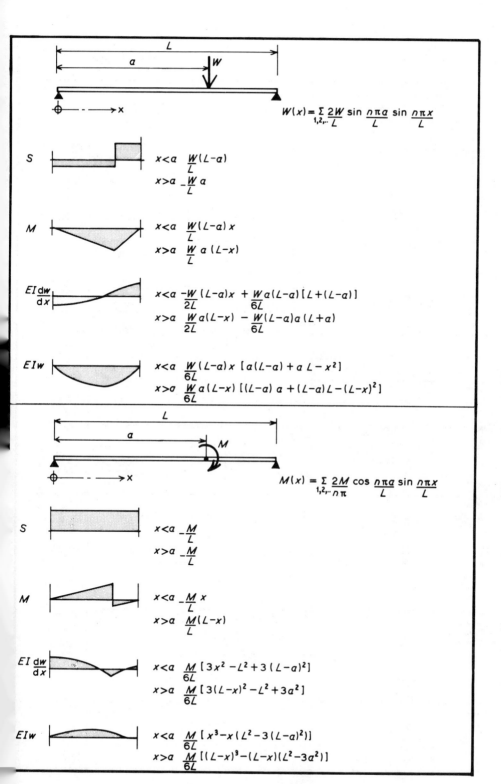

Fig. A.2   Functions of loads on simple span.

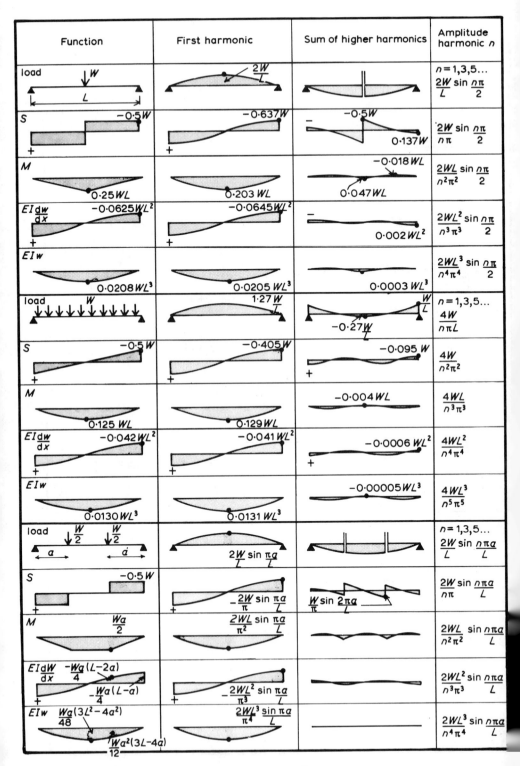

Fig. A.3 Harmonic composition of loads, moments, etc.

APPENDIX B

Approximate folded plate method for beam-and-slab and cellular decks

Notation

Subscripts

0, 1 ...	number of node, beam or web
x, y, z	direction of force, or axis about which moment or torsion rotates or moment of inertia is calculated
b	property of 'beam' section in Fig. B.10.

Superscripts

$'$	relates to top slab of cellular deck
$''$	relates to bottom slab of cellular deck

Symbols

A_1 cross-section area of web 1
A_{b1} cross-section area of 'beam' 1 in Fig. B.10.
A_{01} cross-section area of slab strip 0–1; for cellular deck $A_{01} = l_{01}(d'_{01} + d''_{01})$

$a_{01}, a_{11}, \bar{a}_{01}$	coefficients in stiffness equations
$b_{01}, b_{11}, \bar{b}_{01}, \tilde{b}_{01}$	coefficients in stiffness equations
C_{b1}	torsion constant of 'beam' in Fig. B.10.
$c_{01}, c_{11}, \bar{c}_{01}, c_1$	coefficients in stiffness equations
c_{y01}	transverse torsion constant of slab per unit length; for cellular deck $c_{y01} = c'_{y01} + c''_{y01} = 2i_{01}$
D	D operator = d/dx.
d_{01}	thickness of slab strip 0–1; for cellular deck $d_{01} = d'_{01} + d''_{01}$
d_1	thickness of web 1
E	Young's Modulus
e_{01}, e_{11}	coefficients in stiffness equations
f_{01}, f_{11}	coefficients in stiffness equations
G	Shear Modulus
g_{01}, g_{11}	coefficients in stiffness equations
h_{01}, h_{11}	coefficients in stiffness equations
I_{b1}	moment of inertia of 'beam' of Fig. B.10 about axis Oy
I_{y1}	moment of inertia of web 1 about horizontal axis through centroid
I_{z1}	moment of inertia of web 1 about vertical axis through centroid
I_{z01}	moment of inertia of slab strip 0–1 about vertical axis
I_{01}	$= I^0_{01} + i_{01}l_{01}$ = moment of inertia of slab strips 0–1 about horizontal axis Oy
I^0_{01}	$= A'_{01}z'^2_{01} + A''_{01}z''^2_{01}$
i_1	moment of inertia of web 1 per unit length about longitudinal axis
i_{01}	moment of inertia of slab strip 0–1 per unit length $= d^3_{01}/12$: for cellular decks $i_{01} = i'_{01} + i''_{01}$
j_{01}, j_{11}	coefficients in stiffness equations
k_{01}, k_{11}	coefficients in stiffness equations
L	span between abutments
l_1	height of web 1 between midplanes of top and bottom slabs
l_{01}	width of slab strip 0–1
M_{y1}	bending moment about Oy on cross-section of web 1
M_{z1}	bending moment about Oz on cross-section of web 1
M_{z01}	in-plane bending moment on cross-section of slab strip 0–1
m_{10}	moment on edge 1 of slab strip 0–1 per unit length
m'_1, m''_1	moments on top and bottom edges of web 1
N_1	applied moment on node 1
n	harmonic number

n_{01}, n_{11}	coefficients in stiffness equations
P_1	tensile force on cross-section of web 1
P_{01}	tensile force on cross-section of slab strip 0–1
p_{10}	sideways in-plane force on edge 1 of slab strip 0–1
r_{10}	longitudinal shear force on edge 1 of slab strip 0–1
S_{y1}	sideways shear force on cross-section of web 1
S_{z1}	vertical shear force on cross-section of web 1
s_{10}	shear force downwards on edge 1 of slab strip 0–1
T_1	longitudinal torque about Ox in web 1
t_{10}	transverse torque in slab strip 0–1
u	longitudinal extension of node
v	sideways deflection of node
w	vertical deflection of node
Ox, Oy, Oz	longitudinal, transverse and vertical axes
X_1, Y_1, Z_1	longitudinal, transverse and vertical applied loads on node 1
z'	distance of midplane of top slab *below* Oy
z''	distance of midplane of bottom slab below Oy
z_{b1}	distance of centroid of 'beam' 1 in Fig. B.10 below Oy
$\bar{z}_1$	distance of centroid of web 1 below Oy
α	$= n\pi/L$
ϵ_x	longitudinal tensile strain
ϵ_y	transverse tensile strain
ν	Poisson's Ratio
σ_x	longitudinal tensile stress
σ_y	transverse tensile stress
ϕ_1	sway rotation of node/web 1, clockwise about Ox
ϕ'_1	rotation of top of web 1, clockwise about Ox
ϕ''_1	rotation of bottom of web 1, clockwise about Ox

B.1 Structural limitations and assumptions

This appendix contains the derivation of a set of simple stiffness equations suitable for the analysis of beam-and-slab and cellular bridge decks using a programmable calculator. The equations were used for the derivation of the load distribution charts of Chapter 10.

The method is limited to decks which have:
(1) Simple supports.
(2) Prismatic structure with all slabs horizontal and all webs vertical, as shown in Fig. B.1.

250 *Bridge Deck Behaviour*

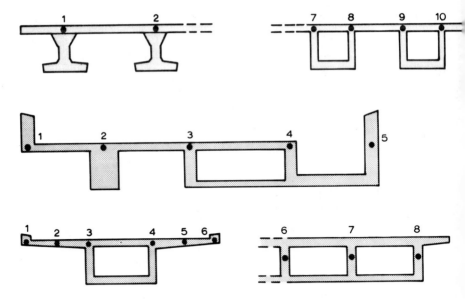

Fig. B.1   Cross-sections of prismatic decks with orthogonal elements.

(3) Right end supports, and at each end a right vertical diaphragm which prevents displacements of the deck in the plane of the diaphragm.

The structural elements are assumed to behave in the following ways:

(4) Beams and webs deflect vertically and compress in accordance with simple beam theory. Shear deformation of webs has been found to be relatively insignificant for conventional design loads, and is ignored.

(5) Horizontal slabs deflect sideways, distort and compress in-plane in accordance with simple beam theory. Shear deformation is considered with plane sections assumed to remain plane.

(6) Transverse flexure of slabs out-of-plane is 'one way'. Transverse moments have component due to Poisson's Ratio interaction of longitudinal flexure.

(7) Longitudinal moments and torques in slabs are 'lumped' with longitudinal moments and torques of adjacent beam or web.

The stiffness equations relate the amplitudes of the nth harmonic component of loads and deflections of the beams or webs. Fig. B.2 shows a plan of the deck over a length equal to the half wavelength of the nth harmonic. It is usually found that for most practical simply supported decks, the harmonics higher than the first harmonic distribute little from the loaded beam. Consequently, the equations need then only be used for the first harmonic, for which the span in Fig. B.2 is the span L between abutments. The forces and deflections are calculated for longitudinal nodes 1, 2, 3, . . . corresponding to the beams and

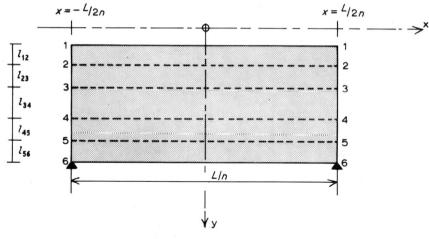

Fig. B.2   Plan of deck.

webs. For slab decks the node spacing is arbitrary. However the approximations in the method lead to errors when the half wavelength is less than three times the node spacing l. Consequently, if information is required for a position in a slab deck near a concentrated load, the node spacing locally should not exceed one third of the half wavelength of the highest harmonic required.

The subsequent analysis considers a cellular structure with more than one slab. Beam-and-slab decks can be considered as cellular decks with bottom slabs of zero thickness.

B.2 Equilibrium of web

Fig. B.3 shows the forces acting on an element of web 1 of a cellular deck. The direction Ox is along the span, Oy transverse and Oz vertically downwards. The origin in this appendix is located at midspan. The stiffness equations derived relate the harmonic amplitudes of the various forces and displacements, and these are not affected by the origin position. The reference nodes are assumed to lie in the Oxy plane which is not necessarily (here) at the same level as the deck. The web centroid is $\bar{z}_1$ below the node. Fig. B.3a shows the applied forces which act on the node; X_1, Y_1 and Z_1 are forces in directions of axes, N_1 is a transverse moment. The following analysis ignores loading between nodes. Consequently Z_1 includes the fixed edge shear forces due to loads on adjacent slabs and N_1 includes the fixed edge moments. (b) shows the forces on the web from interaction with adjacent slabs. The superscript $'$ denotes the top slab while $''$ denotes the bottom slab. The forces m and s result from the transverse slab

252 *Bridge Deck Behaviour*

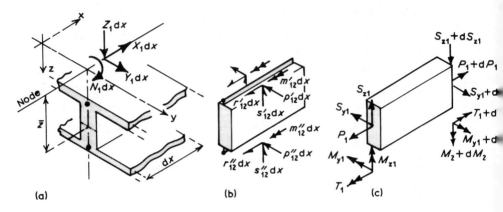

Fig. B.3 Forces on element of web. (a) Applied loads (b) reactions with slabs (c) forces on ends of web element.

flexure, while p and r result from in-plane distortion of the slabs. (c) shows the forces crossing a section of the web, and these include moments and shear forces for biaxial bending, tension P_1 and torsion T_1.

The deflections of the node are shown in Fig. B.4. For the moment it is assumed that the web is 'thick' and does not flex out-of-plane vertically.

All the force and displacement variables are sine or cosine functions. With the

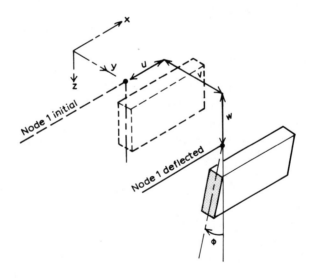

Fig. B.4 Displacements of thick web.

origin at midspan they have the form:

$$
\begin{array}{lll}
Z\cos\alpha x & N\cos\alpha x & Y\cos\alpha x \\
w\cos\alpha x & \phi\cos\alpha x & v\cos\alpha x \\
M\cos\alpha x & P\cos\alpha x & \\
m\cos\alpha x & s\cos\alpha x & p\cos\alpha x \\
& & \\
X\sin\alpha x & S\sin\alpha x & T\sin\alpha x \\
u\sin\alpha x & r\sin\alpha x &
\end{array}
\tag{B.1}
$$

$$\alpha = \frac{n\pi}{L}.$$

In the following discussion, the sin and cos terms are omitted since they are the same throughout an equation. Mention is only made when differentiation takes place.

Equilibrium of the element of web 1 requires

$$
\begin{aligned}
-dS_{z1} + \Sigma s\,dx & = Z_1\,dx \\
-dT_1 + \bar{z}\,dS_{y1} + \Sigma m\,dx - \Sigma pz\,dx & = N_1\,dx \\
-dP_1 + \Sigma r\,dx & = X_1\,dx \\
-dS_{y1} + \Sigma p\,dx & = Y_1\,dx \\
-dM_{z1} - S_{y1}\,dx & = 0 \\
-dM_{y1} + S_{z1}\,dx - \bar{z}_1\,dP_1 + \Sigma rz\,dx & = 0
\end{aligned}
\tag{B.2}
$$

where Σ indicates the sum of forces from all the slabs 0–1 and 1–2 adjoining web 1. These equations can be reduced by differentiation and substitution to

$$
\begin{aligned}
-D^2 M_{y1} - \bar{z}_1 D^2 P_1 + \Sigma s + \Sigma z Dr & = Z_1 \\
-DT_1 - \bar{z}_1 D^2 M_{z1} + \Sigma m - \Sigma zp & = N_1 \\
-DP_1 + \Sigma r & = X_1 \\
D^2 M_{z1} + \Sigma p & = Y_1.
\end{aligned}
\tag{B.3}
$$

B.3 Slab out-of-plane flexure

Fig. B.5 shows the transverse out-of-plane flexure of slab 0–1. We can write the stiffness equations for slabs 0–1 and 1–2 at node 1 directly from Equation 2.10, ignoring the fixed edge forces which are included in Z_1 and N_1. And since for all slabs

$$w' = w'' = w_1$$
$$\phi' = \phi'' = \phi_1$$

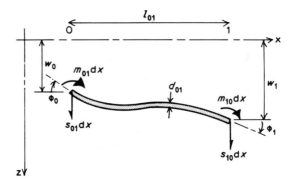

Fig. B.5 Transverse flexure of element of slab.

we can sum the stiffnesses to give

$$\Sigma s = a_{01}w_0 + a_{11}w_1 + a_{12}w_2 - b_{01}\phi_0 + b_{11}\phi_1 + b_{12}\phi_2$$
$$\Sigma m = b_{01}w_0 + b_{11}w_1 - b_{12}w_2 + c_{01}\phi_0 + c_{11}\phi_1 + c_{12}\phi_2 \qquad (B.4)$$

where the coefficients are now

$$a_{01} = -\frac{12Ei_{01}}{(1-\nu^2)l_{01}^3} \qquad b_{01} = \frac{6Ei_{01}}{(1-\nu^2)l_{01}^2} \qquad c_{01} = \frac{2Ei_{01}}{(1-\nu^2)l_{01}}$$

$$a_{11} = -a_{01} - a_{12} \qquad b_{11} = -b_{01} + b_{12} \qquad (B.5)$$

$$c_{11} = c_{01} + c_{12} + \frac{2E}{(1-\nu^2)}\left[\frac{i_{01}}{l_{01}} + \frac{i_{12}}{l_{12}}\right]$$

and i_{01} = sum of slab inertias per unit length

$$= i'_{01} + i''_{01} = \frac{d'^{3}_{01}}{12} + \frac{d''^{3}_{01}}{12}.$$

Interaction with longitudinal flexure is considered later.

Fig. B.6 shows the torsion of a transverse strip of slab under action of torque. If the transverse torsion constant is c_{y01}

$$t_{01}dx = -t_{10}dx = Gc_{y01}dx \left(\frac{dw_1}{dx} - \frac{dw_0}{dx}\right)\frac{1}{l_{01}}$$

or

$$t_{01} = -t_{10} = -\frac{Gc_{y01}\alpha}{l_{01}}(w_1 - w_0)\sin\alpha x$$

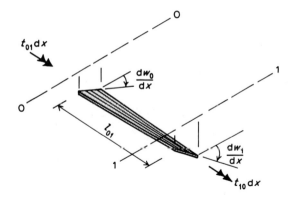

Fig. B.6 Torsion of element of slab.

This torsion can be represented by a statically equivalent shear force on the edge

$$\Delta s_{01} = \frac{dt_{01}}{dx} = -\frac{Gc_{y01}\alpha^2}{l_{01}}(w_1 - w_0) = -\Delta s_{01}. \tag{B.6}$$

Adding this shear force to Equations B.4 for all slabs modifies coefficient a_{01} to

$$a_{01} = -\frac{12Ei_{01}}{(1-\nu^2)l_{01}^3} - \frac{Gc_{y01}\alpha^2}{l_{01}}. \tag{B.7}$$

For a slab

$$c_{y01} = \frac{d_{01}^3}{6} = 2i_{01}.$$

B.4 Web stiffness equations

Since the centroid of the web is located $\bar{z}_1$ below the node, its deflections are

$$u = u_1 - \bar{z}_1 \frac{dw_1}{dx} = (u_1 + \bar{z}_1 \alpha w_1) \sin \alpha x$$

$$v = (v_1 - \bar{z}_1 \phi_1) \cos \alpha x$$

$$w = w_1 \cos \alpha x$$

$$\phi = \phi_1 \cos \alpha x.$$

The torsion stiffness equation for the web is

$$T_1 = T_1 \sin \alpha x = GC_{b1} D(\phi_1 \cos \alpha x) = -GC_{b1}\alpha\phi_1 \sin \alpha x$$
$$DT_1 = -GC_{b1}\alpha^2\phi_1 \cos \alpha x. \tag{B.8}$$

Since the torsion stiffness of adjacent halves of slabs is lumped with the web, C_{b1} is the St. Venant torsion constant of the web with adjacent half slabs (i.e. of the sections in Fig. B.10).

Vertical bending of the web with inertia I_{y1} gives

$$M_{y1} = -EI_{y1}D^2(w_1 \cos \alpha x) = EI_{y1}\alpha^2 w_1 \cos \alpha x$$
$$D^2 M_{y1} = -EI_{y1}\alpha^4 w_1 \cos \alpha x \qquad (B.9)$$

Sideways bending of the web with inertia I_{z1} gives

$$M_{z1} = EI_{z1}D^2(v \cos \alpha x) = -EI_{z1}\alpha^2(v_1 - \bar{z}_1 \phi_1) \cos \alpha x$$
$$D^2 M_{z1} = EI_{z1}\alpha^4(v_1 - \bar{z}_1 \phi_1) \cos \alpha x. \qquad (B.10)$$

Stretching of the web of area A_1 gives

$$P_1 = EA_1 D(u \sin \alpha x) = EA_1(\alpha u_1 + \bar{z}_1 \alpha^2 w_1) \cos \alpha x$$
$$DP_1 = -EA_1(\alpha^2 u_1 + \bar{z}_1 \alpha^3 w_1) \sin \alpha x. \qquad (B.11)$$

Combining Equation B.9 with $\bar{z}_1$ times differential of Equation B.11

$$D^2 M_{y1} + \bar{z}_1 D^2 P_1 = -E(I_{y1} + A_1 z_1^2)\alpha^4 w_1 \cos \alpha x - EA_1 \bar{z}_1 \alpha^3 u_1 \cos \alpha x. \qquad (B.12)$$

B.5 Interaction of longitudinal and transverse moments.

The average deflection of slab 0–1 is

$$w = \frac{(w_0 + w_1)}{2} \cos \alpha x$$

and the average longitudinal curvature

$$\frac{d^2 w}{dx^2} = -\frac{(w_0 + w_1)}{2} \alpha^2 \cos \alpha x$$

hence the modification to transverse moments is

$$\Delta m_{01} = -\Delta m_{10} = -\frac{\nu E i_{01}}{(1-\nu^2)} \frac{d^2 w}{dx^2} = \frac{\nu E i_{01} \alpha^2}{2(1-\nu^2)}(w_0 + w_1)$$

and the additional summed moments on the web are

$$\Delta \Sigma m = \bar{b}_{01} w_0 + \left[-\bar{b}_{01} + \bar{b}_{12} - \frac{\nu E \alpha^2}{(1-\nu^2)}(i_{01} - i_{12}) \right] w_1 - \bar{b}_{12} w_2$$

$$\bar{b}_{01} = -\frac{\nu E \alpha^2 i_{01}}{2(1-\nu^2)}. \qquad (B.13)$$

The average transverse curvature of slab 0–1 is $(\phi_0 - \phi_1)/l_{01}$. Hence the modification to longitudinal moments is

$$\Delta M_{y1} = \frac{\nu E i_{01}}{(1-\nu^2)} \frac{l_{01}}{2} \left(\frac{\phi_0 - \phi_1}{l_{01}}\right) + \frac{\nu E i_{12}}{(1-\nu^2)} \frac{l_{12}}{2} \left(\frac{\phi_1 - \phi_2}{l_{12}}\right).$$

Differentiating twice,

$$\Delta D^2 M_{y1} = -\bar{b}_{01}\phi_0 + \left[-\bar{b}_{01} + \bar{b}_{12} - \frac{\nu E \alpha^2}{(1-\nu^2)}(l_{01} - l_{12})\right]\phi_1 + \bar{b}_{12}\phi_2.$$

(B.14)

B.6 Slab in-plane action

Fig. B.7 shows plan views of the whole of slab strip 0–1, with edge deflections shown in (a) and edge forces in (b). The symmetric and antisymmetric components of these diagrams are shown in Figs. B.8 and B.9 and are considered separately. The forces P_{01}, S_{01} and M_{01} are the in-plane forces crossing a section of the slab strip.

Symmetric in-plane behaviour:
Resolving in the x-direction in Fig. B.8

$$dP_{01} = -(r_{01} + r_{10})dx \sin \alpha x$$

whence average longitudinal tensile stress is

$$\sigma_x = \frac{P_{01}}{l_{01}d_{01}} = \frac{r_{01} + r_{10}}{l_{01}d_{01}\alpha} \cos \alpha x.$$

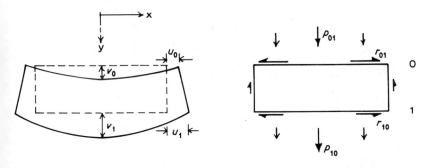

Fig. B.7 In-plane displacements and forces on strip of slab.

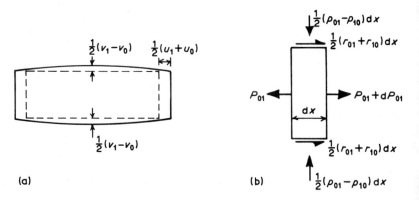

Fig. B.8 Symmetric components of slab in-plane actions. (a) Displacements (b) forces.

The transverse tensile stress is

$$\sigma_y = -\frac{1}{2}\frac{p_{01}-p_{10}}{d_{01}}\cos\alpha x.$$

From the deflections of Fig. B.8 the strains are

$$\epsilon_x = \frac{\partial u}{\partial x} = \tfrac{1}{2}(u_0 + u_1)\alpha\cos\alpha x$$

$$\epsilon_y = \frac{\partial v}{\partial y} = \frac{v_1 - v_0}{l_{01}}\cos\alpha x.$$

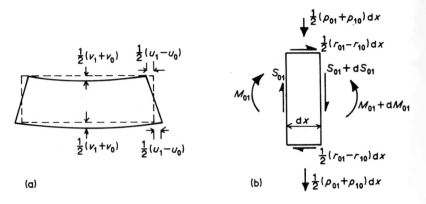

Fig. B.9 Antisymmetric components of slab in-plane actions. (a) Displacements (b) forces.

Plane stress requires

$$\sigma_x = \frac{E}{(1-\nu^2)}\epsilon_x + \frac{\nu E}{(1-\nu^2)}\epsilon_y$$

$$\sigma_y = \frac{E}{(1-\nu^2)}\epsilon_y + \frac{\nu E}{(1-\nu^2)}\epsilon_x$$

whence

$$r_{01} + r_{10} = \frac{El_{01}d_{01}\alpha^2}{(1-\nu^2)2}(u_0 + u_1) + \frac{\nu Ed_{01}\alpha}{(1-\nu^2)}(v_1 - v_0) \quad (B.15)$$

$$p_{01} - p_{10} = -\frac{\nu Ed_{01}\alpha}{(1-\nu^2)}(u_0 + u_1) - \frac{2Ed_{01}}{(1-\nu^2)l_{01}}(v_1 - v_0). \quad (B.16)$$

Antisymmetric in-plane behaviour
Resolving in the y-direction in Fig. B.9.

$$dS_{01} = -(p_{01} + p_{10})dx \cos\alpha x$$

and taking moments

$$dM_{01} = S_{01}dx + \tfrac{1}{2}(r_{01} - r_{10})l_{01}dx \sin\alpha x$$

$$= -\frac{(p_{01} + p_{10})}{\alpha}dx \sin\alpha x + \tfrac{1}{2}(r_{01} - r_{10})l_{01}dx \sin\alpha x.$$

On integrating three times and dividing by $-EI_{z01}$ where $I_{z01} = d_{01}l_{01}^3/12$ we obtain deflection due to bending

$$\tfrac{1}{2}(v_0 + v_1)_{bending} = \left[\frac{(p_{01} + p_{10})}{\alpha^4 I_{z01}E} - \frac{(r_{01} - r_{10})l}{2\alpha^3 EI_{z01}}\right]\cos\alpha x \quad (B.17)$$

which causes longitudinal end displacements

$$\tfrac{1}{2}(u_1 - u_0) = -\frac{l_{01}}{2}\left(\frac{\partial v}{\partial x}\right)_{bending}$$

$$= \frac{l_{01}}{2}\frac{(v_0 + v_1)_{bending}}{2}\alpha\sin\alpha x.$$

Hence

$$\tfrac{1}{2}(u_1 - u_0) = \frac{(p_{01} + p_{10})l_{01}}{2\alpha^3 I_{z01}E} - \frac{(r_{01} - r_{10})l_{01}^2}{4\alpha^2 I_{z01}E}. \quad (B.18)$$

There is also a component of sideways deflection due to shear deformation. Assuming plane sections remain plane

$$\left(\frac{\partial v}{\partial x}\right)_{shear} = \frac{S_{01}}{Gl_{01}d_{01}} = -\frac{(p_{01}+p_{10})}{Gl_{01}d_{01}\alpha}\sin \alpha x.$$

Hence

$$\tfrac{1}{2}(v_0+v_1)_{shear} = \frac{(p_{01}+p_{10})}{Gl_{01}d_{01}\alpha^2}\cos \alpha x. \tag{B.19}$$

Combining Equations B.17 and B.19,

$$\tfrac{1}{2}(v_0+v_1) = (p_{01}+p_{10})\left[\frac{1}{\alpha^4 I_{z01}E} + \frac{1}{\alpha^2 l_{01}d_{01}G}\right] - \frac{(r_{01}-r_{10})l_{01}}{2\alpha^3 I_{z01}E}. \tag{B.20}$$

Rearranging Equations B.18 and B.20, and writing $d_{01}l_{01}^3/12$ for I_{z01} and $E/2(1+\nu)$ for G we obtain

$$(p_{01}+p_{10}) = \frac{\alpha^2 l_{01}d_{01}E}{4(1+\nu)}(v_0+v_1) - \frac{\alpha d_{01}E}{2(1+\nu)}(u_1-u_0) \tag{B.21}$$

$$(r_{01}-r_{10}) = \frac{\alpha d_{01}E}{2(1+\nu)}(v_0+v_1) - \left[\frac{\alpha^2 l_{01}d_{01}E}{6} + \frac{d_{01}E}{(1+\nu)l_{01}}\right](u_1-u_0) \tag{B.22}$$

From Equations B.15, B.16, B.21 and B.22,

$$r_{10} = g_{01}u_0 + \left[-g_{01} + \frac{\alpha^2 l_{01}d_{01}E}{2(1-\nu^2)}\right]u_1 - k_{01}v_0 + \left[-k_{01} + \frac{\nu\alpha d_{01}E}{(1-\nu^2)}\right]v_1 \tag{B.23}$$

$$p_{10} = k_{01}u_0 + \left[-k_{01} + \frac{\nu\alpha d_{01}E}{(1-\nu^2)}\right]u_1 + n_{01}v_0 + \left[n_{01} + \frac{2d_{01}E}{(1-\nu^2)l_{01}}\right]v_1 \tag{B.24}$$

where

$$g_{01} = -\left[\frac{1}{2(1+\nu)l_{01}^2} - \alpha^2\left(\frac{1}{4(1-\nu^2)} - \frac{1}{12}\right)\right]A_{01}E$$

$$k_{01} = \frac{\alpha d_{01}E}{4(1-\nu)}$$

$$n_{01} = -\left[\frac{1}{(1-\nu^2)l_{01}^2} - \frac{\alpha^2}{8(1+\nu)}\right]A_{01}E \tag{B.25}$$

$$A_{01} = l_{01}d_{01}.$$

Appendix B   261

At the level of the top slab in the structure the displacements u and v are related to node deflections by

$$u' = u_1 + z'\alpha w_1 \qquad v' = v_1 - z'\phi_1$$

with similar equations for the bottom slab. Making this substitution in Equations B.23 and B.24 and summing all similar equations for slab edges attached to web 1, we obtain

$$\Sigma r = g_{01}u_0 + \left[-g_{01} - g_{12} + \frac{\alpha^2 E}{(1-\nu^2)}\left(\frac{A_{01} + A_{12}}{2}\right)\right]u_1 + g_{12}u_2$$

$$+ e_{01}w_0 + \left[-e_{01} - e_{12} + \frac{\alpha^3 E}{(1-\nu^2)}\left(\frac{A_{01}z_{01} + A_{12}z_{12}}{2}\right)\right]w_1 + e_{12}w_2$$

$$- k_{01}v_0 + \left[-k_{01} + k_{12} + \frac{\nu\alpha E}{(1-\nu^2)}(d_{01} - d_{12})\right]v_1 + k_{12}v_2$$

$$+ f_{01}\phi_0 + \left[f_{01} - f_{12} - \frac{\nu\alpha E}{(1-\nu^2)}(d_{01}z_{01} - d_{12}z_{12})\right]\phi_1 - f_{12}\phi_2$$

(B.26)

and

$$\Sigma p = k_{01}u_0 + \left[-k_{01} + k_{12} + \frac{\nu\alpha E}{(1-\nu^2)}(d_{01} - d_{12})\right]u_1 - k_{12}u_2$$

$$+ h_{01}w_0 + \left[-h_{01} + h_{12} + \frac{\nu\alpha^2 E}{(1-\nu^2)}(d_{01}z_{01} - d_{12}z_{12})\right]w_1 - h_{12}w_2$$

$$+ n_{01}v_0 + \left[n_{01} + n_{12} + \frac{2E}{(1-\nu^2)}\left(\frac{d_{01}}{l_{01}} + \frac{d_{12}}{l_{12}}\right)\right]v_1 + n_{12}v_2$$

$$+ j_{01}\phi_0 + \left[j_{01} + j_{12} - \frac{2E}{(1-\nu^2)}\left(\frac{d_{01}z_{01}}{l_{01}} + \frac{d_{12}z_{12}}{l_{12}}\right)\right]\phi_1 + j_{12}\phi_2.$$

(B.27)

The expressions for g, k and n are identical to Equation B.25 except that now the symbols are equal to the sum of plate properties.
d_{01} = sum of slab thicknesses = $d'_{01} + d''_{01}$
A_{01} = total slab cross-section area = $l_{01}(d'_{01} + d''_{01})$
and the additional coefficients in Equations B.26 and B.27 have values

$$e_{01} = z_{01}\alpha g_{01} \qquad f_{01} = z_{01}k_{01}$$
$$h_{01} = z_{01}\alpha k_{01} \qquad j_{01} = -z_{01}n_{01}$$

(B.28)

where z_{01} = level of centroid of slabs

$$= \frac{z_{01}'d_{01}' + z_{01}''d_{01}''}{d_{01}' + d_{01}''}.$$

To obtain $\Sigma z Dr$, all terms of Equation B.26 are multiplied by z' or z'' and differentiated by further multiplication by α. Hence

$$\begin{aligned}
\Sigma z Dr = {} & +e_{01}u_0 + \left[-e_{01} - e_{12} + \frac{\alpha^3 E}{(1-\nu^2)}\left(\frac{A_{01}z_{01} + A_{12}z_{12}}{2}\right)\right]u_1 + e_{12}u_2 \\
& + \bar{a}_{01}w_0 + \left[-\bar{a}_{01} - \bar{a}_{12} + \frac{\alpha^4 E}{(1-\nu^2)}\left(\frac{I_{01}^0 + I_{12}^0}{2}\right)\right]w_1 + \bar{a}_{12}w_2 \\
& - h_{01}v_0 + \left[-h_{01} + h_{12} + \frac{\nu\alpha^2 E}{(1-\nu^2)}(d_{01}z_{01} - d_{12}z_{12})\right]v_1 + h_{12}v_2 \\
& - \bar{b}_{01}\phi_0 + \left[-\bar{b}_{01} + \bar{b}_{12} - \frac{\nu\alpha^2 E}{(1-\nu^2)}\left(\frac{I_{01}^0}{l_{01}} - \frac{I_{12}^0}{l_{12}}\right)\right]\phi_1 + \bar{b}_{12}\phi_2
\end{aligned}$$

(B.29)

where

$$\bar{a}_{01} = -\left[\frac{1}{2(1+\nu)l_{01}^2} - \alpha^2\left(\frac{1}{4(1-\nu^2)} - \frac{1}{12}\right)\right]\alpha^2 I_{01}^0 E$$

$$\bar{b}_{01} = -\frac{\alpha^2 I_{01}^0 E}{4(1-\nu)l_{01}}$$ (B.30)

$$I_{01}^0 = l_{01}(d_{01}'z_{01}'^2 + d_{01}''z_{01}''^2).$$

Multiplying all coefficients of Equation B.27 by $-z'$ or $-z''$ we obtain

$$\begin{aligned}
-\Sigma z p = {} & -f_{01}u_0 + \left[+f_{01} - f_{12} - \frac{\nu\alpha E}{(1-\nu^2)}(d_{01}z_{01} - d_{12}z_{12})\right]u_1 + f_{12}u_2 \\
& + \bar{b}_{01}w_0 + \left[-\bar{b}_{01} + \bar{b}_{12} - \frac{\nu\alpha^2 E}{(1-\nu^2)}\left(\frac{I_{01}^0}{l_{01}} - \frac{I_{12}^0}{l_{12}}\right)\right]w_1 - \bar{b}_{12}w_2 \\
& + j_{01}v_0 + \left[j_{01} + j_{12} - \frac{2E}{(1-\nu^2)}\left(\frac{d_{01}z_{01}}{l_{01}} + \frac{d_{12}z_{12}}{l_{12}}\right)\right]v_1 + j_{12}v_2 \\
& + \bar{c}_{01}\phi_0 + \left[\bar{c}_{01} + \bar{c}_{12} + \frac{2E}{(1-\nu^2)}\left(\frac{I_{01}^0}{l_{01}^2} + \frac{I_{12}^0}{l_{12}^2}\right)\right]\phi_1 + \bar{c}_{12}\phi_2
\end{aligned}$$

(B.31)

where

$$\bar{c}_{01} = -\left[\frac{1}{(1-\nu^2)l_{01}^2} - \frac{\alpha^2}{8(1+\nu)}\right]I_{01}^0 E.$$

B.7 Node stiffness equations

It is now possible to substitute the various functions for Σm, Σr, etc. into the left hand side of Equations B.3 to obtain the stiffness Equations B.32–B.35 for node 1. On combining various terms for slab in-plane action and web bending it is found that considerable simplification results if the various components are summed as section properties of the 'beam' sections in Fig. B.10.

z_{b1} = level of centroid of 'beam' below Oy

I_{b1} = moment of inertia of 'beam' 1 about Oy (note not centroid)

$$= \frac{i_{01}l_{01} + i_{12}l_{12}}{2(1-\nu^2)} + \frac{I_{01}^0 + I_{12}^0}{2(1-\nu^2)} + I_{y1} + A_1\bar{z}_1^2$$

$$A_{b1} = \frac{A_{01} + A_{12}}{2(1-\nu^2)} + A_1.$$

In addition, $I_{01} = I_{01}^0 + i_{01}l_{01}$.

Equation B.32 is obtained from Equations B.12, B.14, B.29 and B.4 (with definitions B.5 and B.7).

Equation B.33 is obtained from Equations B.8, B.10, B.31 and B.4 with B.13.

Equation B.34 is obtained from Equations B.11 and B.26.

Equation B.35 is obtained from Equations B.10 and B.27.

$$\begin{bmatrix} a_{01} & a_{11} & a_{12} & -b_{01} & b_{11} & b_{12} & e_{01} & e_{11} & e_{12} & -h_{01} & h_{11} & h_{12} \\ b_{01} & b_{11} & -b_{12} & c_{01} & c_{11} & c_{12} & -f_{01} & f_{11} & f_{12} & j_{01} & j_{11} & j_{12} \\ e_{01} & e_{11} & e_{12} & f_{01} & f_{11} & -f_{12} & g_{01} & g_{11} & g_{12} & -k_{01} & k_{11} & k_{12} \\ h_{01} & h_{11} & -h_{12} & j_{01} & j_{11} & j_{12} & k_{01} & k_{11} & -k_{12} & n_{01} & n_{11} & n_{12} \end{bmatrix} \begin{bmatrix} w_0 \\ w_1 \\ w_2 \\ \phi_0 \\ \phi_1 \\ \phi_2 \\ u_0 \\ u_1 \\ u_2 \\ v_0 \\ v_1 \\ v_2 \end{bmatrix}$$

$$= \begin{bmatrix} Z_1 \\ N_1 \\ X_1 \\ Y_1 \end{bmatrix}$$

(B.32)
(B.33)
(B.34)
(B.35)

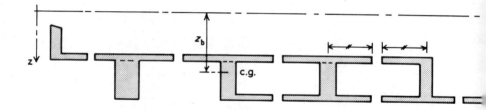

Fig. B.10 Subdivision of deck into 'beams'.

where

$$a_{01} = -\frac{Ec_{y01}\alpha^2}{(1-\nu^2)l_{01}^3} - \frac{Ec_{y01}\alpha^2}{2(1+\nu)l_{01}} - \left[\frac{1}{2(1+\nu)l_{01}^2} - \alpha^2\left(\frac{1}{4(1-\nu^2)} - \frac{1}{12}\right)\right]E\alpha^2$$

$$a_{11} = -a_{01} - a_{12} + E\alpha^4 I_{b1}$$

$$b_{01} = \frac{6Ei_{01}}{(1-\nu^2)l_{01}^2} - \frac{\nu E\alpha^2 i_{10}}{2(1-\nu^2)} - \frac{E\alpha^2 I_{01}^0}{4(1-\nu)l_{01}}$$

$$b_{11} = -b_{01} + b_{12} - \frac{\nu E\alpha^2}{(1-\nu^2)}\left(\frac{I_{01}}{l_{01}} - \frac{I_{12}}{l_{12}}\right)$$

$$c_{01} = \frac{2Ei_{01}}{(1-\nu^2)l_{01}} - \left[\frac{1}{(1-\nu^2)l_{01}^2} - \frac{\alpha^2}{8(1+\nu)}\right]EI_{01}^0$$

$$c_{11} = c_{01} + c_{12} + \frac{2E}{(1-\nu^2)}\left(\frac{I_{01}}{l_{01}^2} + \frac{I_{12}}{l_{12}^2}\right) + E\alpha^4 \bar{z}_1^2 I_{z1} + \frac{E}{2(1+\nu)}\alpha^2 C_{b1}$$

$$e_{01} = \alpha z_{01} g_{01}$$

$$e_{11} = -e_{01} - e_{12} + E\alpha^3 z_{b1} A_{b1}$$

$$f_{01} = z_{01} k_{01}$$

$$f_{11} = f_{01} - f_{12} - \frac{\nu E\alpha}{(1-\nu^2)}(d_{01}z_{01} - d_{12}z_{12})$$

$$g_{01} = -\left[\frac{1}{2(1+\nu)l_{01}^2} - \alpha^2\left(\frac{1}{4(1-\nu^2)} - \frac{1}{12}\right)\right]EA_{01}$$

$$g_{11} = -g_{01} - g_{12} + E\alpha^2 A_{b1}$$

$$h_{01} = \alpha f_{01}$$

$$h_{11} = -\alpha f_{11}$$

$$j_{01} = -z_{01}n_{01}$$

$$j_{11} = j_{01} + j_{12} - \frac{2E}{(1-\nu^2)} \left(\frac{d_{01}z_{01}}{l_{01}} + \frac{d_{12}z_{12}}{l_{12}} \right) - E\alpha^4 \bar{z}_1 I_{z1}$$

$$k_{01} = \frac{E\alpha d_{01}}{4(1-\nu)}$$

$$k_{11} = k_{01} + k_{12} + \frac{\nu E\alpha}{(1-\nu^2)}(d_{01} - d_{12})$$

$$n_{01} = -\left[\frac{1}{(1-\nu^2)l_{01}^2} - \frac{\alpha^2}{8(1+\nu)} \right] EA_{01}$$

$$n_{11} = n_{01} + n_{12} + \frac{2E}{(1-\nu^2)} \left(\frac{d_{01}}{l_{01}} + \frac{d_{12}}{l_{12}} \right) + \alpha^4 I_{z1}.$$

B.8 Node stiffness equations for plane decks

If the depth z_{01} to the centroid of the slabs is the same between all nodes, considerable simplification ensues when the Oy axis is placed at the same level as the centroid plane. For a beam-and-slab deck this would be in the midplane of the slab, while for a cellular deck it would be at the level of the centroid of the two slabs (cantilever slabs must be considered as part of edge webs). Under this condition

$$e_{01} = e_{12} = h_{01} = h_{11} = h_{12} = f_{01} = f_{11} = f_{12} = j_{01} = j_{12} = 0. \quad (B.36)$$

A further considerable simplification ensues if the deck has a cross-section which is symmetric about Oy. For a beam-and-slab deck this means that slab membrane action is ignored in the same way that it is in a grillage. The node stiffness equations can then be written

$$\begin{bmatrix} a_{01} & a_{11} & a_{12} & -b_{01} & b_{11} & b_{12} \\ b_{01} & b_{11} & -b_{12} & c_{01} & c_{11} & c_{12} \end{bmatrix} \begin{bmatrix} w_0 \\ w_1 \\ w_2 \\ \phi_0 \\ \phi_1 \\ \phi_2 \end{bmatrix} = \begin{bmatrix} Z_1 \\ N_1 \end{bmatrix} \quad (B.37)$$

If Poisson's Ratio is $\nu = 0.2$ and second order terms are ignored, these coefficients

are

$$a_{01} = -\left(\frac{12.5i_{01}}{l_{01}^3} + \frac{0.42c_{y01}\alpha^2}{l_{01}} + \frac{0.42\alpha^2 I_{01}^0}{l_{01}^2}\right)E$$

$$a_{11} = -a_{01} - a_{12} + \alpha^4 I_{b1}E$$

$$b_{01} = \left(\frac{6.25i_{01}}{l_{01}^2} - 0.10i_{01}\alpha^2 - \frac{0.31 I_{01}^0}{l_{01}}\right)E$$

$$b_{11} = -b_{01} + b_{12} - 0.21\alpha^2 \left(\frac{I_{01}^0}{l_{01}} - \frac{I_{12}^0}{l_{12}}\right)E \qquad (B.38)$$

$$c_{01} = \left(\frac{2.08i_{01}}{l_{01}} - \frac{1.04 I_{01}^0}{l_{01}^2} + 0.10\alpha^2 I_{01}^0\right)E$$

$$c_{11} = c_{01} + c_{12} + 2.08\left(\frac{I_{01}}{l_{01}^2} + \frac{I_{12}}{l_{12}^2}\right)E + 0.42\alpha^2 C_{b1}E.$$

When the analysis is primarily intended to determine load distribution rather than deflections, computation is reduced if E is set equal to unity.

B.9 Node stiffness equations for thin webbed cellular decks.

Equations B.37 can be modified to accommodate decks with flexible webs. Each web now has three rotational displacements ϕ, ϕ' and ϕ'' as shown in Fig. B.11.

Fig. B.11 Out-of-plane flexure of thin web.

Appendix B 267

It is now necessary to split the moment equation for N_1 in Equation B.3 into three components:

for rotation of the top of the web through ϕ_1' under applied moment N_1', where m_1' is the moment at the top of the web.

$$m_{10}' + m_{12}' + m_1' = N_1' \qquad (B.39)$$

for rotation of bottom of web through ϕ_1'' under the applied moment N_1''.

$$m_{10}'' + m_{12}'' + m_1'' = N_1'' \qquad (B.40)$$

for sway rotation ϕ_1 of the web.

$$-DT_1 - \bar{z}_1 D^2 M_{z1} - \Sigma zp - m_1' - m_1'' = 0 \qquad (B.41)$$

The slope deflection equations for the web are

$$\begin{aligned} m_1' &= \frac{Ei_1}{(1-\nu^2)l_1} (4\phi_1' + 2\phi_1'' - 6\phi_1) \\ m_1'' &= \frac{Ei_1}{(1-\nu^2)l_1} (2\phi_1' + 4\phi_1'' - 6\phi_1) \end{aligned} \qquad (B.42)$$

where i_1 = moment of inertia of web about the longitudinal axis, per unit length = $d_1^3/12$.

Using these Equations we can modify Equations B.37 into Equation B.43

$$\begin{bmatrix} a_{01} & a_{11} & a_{12} & -\bar{b}_{01} & \bar{b}_{11} & \bar{b}_{12} & -b_{01}' & b_{11}' & b_{12}' & -b_{01}'' & b_{11}'' & b_{12}'' \\ \bar{b}_{01} & \bar{b}_{11} & -\bar{b}_{12} & \bar{c}_{01} & \bar{c}_{11} & \bar{c}_{12} & 0 & -3c_1 & 0 & 0 & -3c_1 & 0 \\ b_{01}' & b_{11}' & -b_{12}' & 0 & -3c_1 & 0 & c_{01}' & c_{11}' & c_{12}' & 0 & c_1 & 0 \\ b_{01}'' & b_{11}'' & -b_{12}'' & 0 & -3c_1 & 0 & 0 & c_1 & 0 & c_{01}'' & c_{11}'' & c_{12}'' \end{bmatrix} \begin{bmatrix} w_0 \\ w_1 \\ w_2 \\ \phi_0 \\ \phi_1 \\ \phi_2 \\ \phi_0' \\ \phi_1' \\ \phi_2' \\ \phi_0'' \\ \phi_1'' \\ \phi_2'' \end{bmatrix}$$

$$= \begin{bmatrix} Z_1 \\ 0 \\ N_1' \\ N_1'' \end{bmatrix} \qquad (B.43)$$

where

$$a_{01} = -\frac{12Ei_{01}}{(1-\nu^2)l_{01}^3} - \frac{Ec_{y01}\alpha^2}{2(1+\nu)l_{01}}$$

$$- \left[\frac{1}{2(1+\nu)l_{01}^2} - \alpha^2\left(\frac{1}{4(1-\nu^2)} - \frac{1}{12}\right)\right] E\alpha^2 I_{01}^0$$

$$a_{11} = -a_{01} - a_{12} + E\alpha^4 I_{b1}$$

$$\bar{b}_{01} = -\frac{E\alpha^2 I_{01}^0}{4(1-\nu)l_{01}}$$

$$b_{01}' = \frac{6Ei_{01}'}{(1-\nu^2)l_{01}^2} - \frac{\nu E\alpha^2 i_{01}'}{2(1-\nu^2)}$$

$$b_{01}'' = \frac{6Ei_{01}''}{(1-\nu^2)l_{01}^2} - \frac{\nu E\alpha^2 i_{01}''}{2(1-\nu^2)}$$

$$\bar{b}_{11} = -\bar{b}_{01} + \bar{b}_{12} - \frac{\nu E\alpha^2}{(1-\nu^2)}\left[\frac{I_{01}^0}{l_{01}} - \frac{I_{12}^0}{l_{12}}\right]$$

$$b_{11}' = -b_{01}' + b_{12}' - \frac{\nu E\alpha^2}{(1-\nu^2)}\left[\frac{i_{01}'}{l_{01}} - \frac{i_{12}'}{l_{12}}\right]$$

$$b_{11}'' = -b_{01}'' + b_{12}'' - \frac{\nu E\alpha^2}{(1-\nu^2)}\left[\frac{i_{01}''}{l_{01}} - \frac{i_{12}''}{l_{12}}\right]$$

$$\bar{c}_{01} = -\left[\frac{1}{(1-\nu^2)l_{01}^2} - \frac{\alpha^2}{8(1+\nu)}\right]EI_{01}^0$$

$$\bar{c}_{11} = \bar{c}_{01} + \bar{c}_{12} + \frac{2E}{(1-\nu^2)}\left[\frac{I_{01}^0}{l_{01}^2} + \frac{I_{12}^0}{l_{12}^2}\right] + \frac{E\alpha^2 C_{b1}}{2(1+\nu)} + 6c_1$$

$$c_{01}' = \frac{2Ei_{01}'}{(1-\nu^2)l_{01}} \quad c_{01}'' = \frac{2Ei_{01}''}{(1-\nu^2)l_{01}} \quad c_1 = \frac{2Ei_1}{(1-\nu^2)l_1}$$

$$c_{11}' = 2c_{01}' + 2c_{12}' + 2c_1 \quad c_{11}'' = 2c_{01}'' + 2c_{12}'' + 2c_1.$$

Author index

Abdel-Samad. S. R., 115
Abel, J. F., 242
American Institute of Steel Construction, 68
Armer, G. S. T., 162

Balas, J., 162, 185
Bannister, J. L., 242
Beckett, D., 18
Bell, K., 242
van den Berg, J., 162
Best, B. C., 125
Blaszkowiak, S., 162
British Railways Board, 141

Case, J. 45
Chaudhuri, B. K., 162
Cheung, Y. K., 242
Chilver, A. H., 45
Clark, L. A., 162
Clough, R. W., 242
Coates, R. C., 45
Coutie, M. G., 45
Cusens, A. R., 185

Department of Environment, 125, 152, 185, 186
Desai, C. S., 242

Evans, H. R., 242

DeFries-Skene, A., 223

Goldberg, J. E., 223
Goodier, J. N., 45

Hanuska, A., 162, 185
Harris, F. R., 162
Harris, J. D., 203
Hendry, A. W., 223
Hergenroder, A., 162, 185
Holand, I., 242
Hook, D. M. A., 141

Jaeger, L. G., 68, 223

Kaczkowski, Z., 162
Kong, F. K., 45
Kreyszig, E., 223

Lampert, P., 86
Leve, H. L., 223
Libby, J. R., 203
Lightfoot, E., 45, 69
Little, G., 185

McHenry, D., 141
Maisel, B. I., 115
Martin, H. C., 242
Morice, P. B., 45, 185

Oden, J. T., 45

Pama, R. P., 185
Pucher, A., 69, 86, 115, 185, 223

Richmond, B., 141
Robinson, A. R., 115
Rockey, K. C., 242
Rowe, R. E., 68, 86, 185, 223
Rusch, H., 162, 185

270 *Index*

Sawko, F., 69, 115, 186
Scordelis, A. C., 223
Smith, I. C., 203
Spindel, J. E., 125

Timoshenko, S., 45
Topp, L. J., 242
Toppler, J. F., 162
Troitsky, M. S., 68

Turner, M. J., 242

West, R., 69, 85
Westergaard, H. M., 223
Witecki, A. A., 162
Wood, R. H., 162
Wright, R. N., 115

Zienkiewicz, O. C., 242

Subject index

Antisymmetric components, 207
Articulated plate theory, 116
Articulation, 20
Axes, 21
Axial compression of beams, 131

Beam
    analysis, 19–45
    box, 7
    finite element, 236
    inverted T, 11
Beam decks, 4, 19–45
Beam on elastic foundations, 114
'Beam', 166
Beam-and-slab deck, 9, 70–86
    box beams, 12, 71, 80
    charts, 166–181
    contiguous beam, 11, 70, 74
    downstand grillage, 127
    effective flange, 140, 143
    equilibrium equations, 73
    finite elements, 237
    grillage, 74
    skew, 155
    spaced beam, 11, 70, 74
    spine beam, 13
    stiffnesses, 74
Bending
    beam, 21, 225
    cellular deck, 90, 94, 109
    in-plane, 127
    moment charts, 167
    moment diagram, 23, 67, 83, 110, 245, 246
    stress distribution, 23, 133, 144
    transverse, 94, 111
    'true' moments, 67, 110

Box beam
    contiguous, 7, 13
    edge beam, 146
    chart analysis, 177
    shear key deck, 122
    skew deck, 155
Box girder, 4, 17

Cantilever construction, 13, 17
Cellular deck, 13, 87–115
    charts, 181
    effective flange, 143
    finite element, 231, 237
    folded plate, 216
    shear lag, 144
    temperature, 195
Cellular stiffness ratio, 166, 182
Chart analysis, 163–186
    edge 'beam', 167
    internal 'beam', 169
    next-to-edge 'beam', 168
Complementary shear stresses, 25
Complementary torques, 49
Composite construction, 9, 11, 17, 47, 65, 77, 79
Concentrated load effects, 68, 72, 82, 109, 112, 212
Concrete, 7, 60, 77, 196
Conforming element, 235, 237
Construction sequence, 43
Contiguous beams, 11, 70, 74
Continuous beam analysis, 19–45
    cellular deck, 114
    computer, 42
    right wide decks, 21, 114
Continuous skew decks, 155, 156
Coordinates, local and global, 218

Index 271

Creep, 45, 196
Cruciform space frame, 137
Curved beam slope-deflection, 161
Curved deck, 159–162
  beam, 20
  beam-and-slab, 71
  equilibrium, 159
  finite strip, 240
  grillage, 161
Cylindrical voids, 7, 13, 65, 88, 99

Dead load, 43
Determinate structure, 19, 37
Diaphragm, 4, 10, 14, 70, 94, 100, 137
  folded plate structure, 217, 239
  skew deck, 157
Displacement field, 226
Displacement method (see flexibility coefficients)
Distortion of cross-section (see also shear flexibility), 4, 17, 90, 97, 100, 111, 122, 165
Distortion of flange (see shear lag)
Distribution of harmonics, 210
Dowels, 119
Downstand grillage, 127

Edge beam and edge stiffening, 48, 119, 146–152, 154
Effective width of flange, 76, 77, 93, 140, 142–147

Finite element analysis, 7, 17, 46, 114, 145, 185, 197, 224–242
Finite strip analysis, 238
First harmonic, 165, 172
Fixed end moments and shear forces, 27, 81, 109
Flange, effective width, 142–147
Flexibility coefficients, 31, 201
Flexural stiffness ratio, 166, 174, 182
Folded plate analysis, 17, 114, 148, 166, 204–223, 247–268
Force method (see stiffness equations and coefficients)
Fourier analysis (see harmonic analysis)

Gibb's phenomenon, 222, 240
Grid deck, 4, 71
Grillage analysis
  beam-and-slab deck, 13, 74–86
  curved deck, 161
  mesh and member spacing, 57, 74, 88, 119, 181
  output interpretation, 67, 82, 109
  parapets, 148
  shear flexible, 17, 87–115, 121
  shear key deck, 119–125

  skew deck, 156
  skew mesh, 58, 156
  slab deck, 7, 55–69
  tapered deck, 158
  temperature effects, 195
  torsion, 57, 61, 76, 95
Grillage examples
  beam-and-slab deck, 77, 79, 80
  cellular deck, 103, 106, 107
  grid deck, 75
  shear key deck, 123
  slab deck, 63, 64, 65
Grillage section properties
  beam-and-slab deck, 75
  cellular deck, 93
  shear key deck, 122
  skew deck, 157
  slab deck, 59

Harmonic analysis, 165, 204–223, 238, 245
Harmonic components, 245, 246
Haunches, 28, 88, 107, 192, 198

Inclined webs, 87, 107, 137
Influence coefficients (see flexibility coefficients)
Influence lines, 43, 172
Influence value, 171
In-plane forces and displacements, 83, 126, 142, 197, 216, 225
Isotropic slab, 7, 47, 164

Loading, 68, 81, 108
Local moments (see concentrated load effects)
Longitudinal torsion, 100
Lower bound, 233, 236

McHenry lattice, 134
Membrane action (see in-plane forces and displacements)
Membrane analogy, 41
Moment-curvature equations, 26, 50
Moment of inertia, 24, 59, 75, 93
Multicellular decks (see cellular decks)
Multiple span beam, 35

Neutral axis, 23, 84, 91, 139, 147–152
Non-conforming element, 235–238

Orthotropic slab, 7, 48, 64, 164
Out-of-plane forces and deflections (see also bending), 216

Parapets, 148
Plane frame analysis, 12, 81, 98
Plane stress (see also in-plane forces and displacements), 225

Plate bending, 48–52, 233
Poisson's Ratio, 56, 94, 135, 137, 194, 237
Prestress, 44, 77, 196–203
Principal moments, 53
Principal stresses, 54
Prismatic decks, 204
Product integral, 33, 203, 244

Quadrilateral element, 232, 235

Raking piers, 45
Reciprocal theorem, 172
Redistribution of moments, 45, 85
Reliability, 241
Rotation of 'beam', 172
Rotational stiffness ratio, 166, 174, 182

Saint-Venant, 39, 211
Second moment of area (see moment of inertia)
Section properties (see grillage section properties)
Service bay, 150
Shear area, 28, 98, 121
Shear centre, 36
Shear deformation (see also shear flexibility), 126
Shear flexibility (see also grillage analysis), 28, 75, 97, 121, 138
Shear force
   bending, 22, 49, 67, 110
   diagram, 23, 245, 246
   distribution, 172, 213
   torsion, 49, 95, 112
   web, 96
Shear flow, 41, 51, 57, 61, 91, 93, 95, 142
   flange, 126
   inter-beam, 126
Shear key deck, 9, 116–125
   charts, 116, 166
Shear lag, 142–147
Shear Modulus, 28, 39
Shear stress, 24, 41, 51
Shell finite element, 237
Shrinkage, 196
Simply supported beam functions, 245, 246
Skew deck, 11, 153–158
   beam, 20
   beam-and-slab, 77
   cellular, 103, 106
   folded plate, 221
   shear key, 124
   slab, 64, 65
Skew grillage (see grillage analysis)
Slab deck, 7, 46–69
   charts, 163–177
   composite, 7, 47, 64
   edge stiffening, 146
   finite element, 7, 233
   isotropic, 7, 47, 164
   orthotropic, 7, 48, 64, 164
   voided, 7, 47, 65
Slab membrane action (see in-plane forces and displacements)
Slopes and deflections, 25
Slope-deflection equations, 27, 161
Space frame analysis, 17, 126–141
Statical distribution, 81, 109, 212
Steel, 11, 17, 79, 144
Stiffness equations and coefficients, 29–32, 215, 217, 228, 236
Strain, longitudinal, 22, 58
Supports, 19, 125, 153, 221
Symmetric components, 207

Tapered deck, 158
Temperature effects, 187–196
Three-dimensional finite elements, 237–241
Three-dimensional space frame, 126–141
Torque-twist equation, 38
Torsion
   beam, 36–42
   beam-and-slab, 76, 159
   cellular deck, 13, 95
   grillage, 57, 61, 95
   longitudinal, 100
   saw teeth, 67, 109
   shear key deck, 123
   skew deck, 153
   slab deck, 51
   transverse, 100
   warping, 100, 131
Torsion constant
   beam, 40
   cellular deck, 42, 95
   concrete, 77
   grillage member, 61, 76, 95, 123
   slab, 51, 61
Torsional rigidity, 39
Transverse beams, 73, 77
Trapezoidal cell, 107
Trapezoidal shear, 128
Triangular finite element, 225, 233
Trigonometric analysis (see harmonic analysis)

U-beam, 71, 152
Uplift, 153
Upper bound, 233, 236

Voided slab, 7, 47, 65

Warping, 84, 100, 131
Wheel loads (see concentrated load effects)

Young's Modulus, 23, 50